Professional Ethics for the Construction Industry

Construction professionals have a range of demanding responsibilities: towards clients, their companies, and to abide by government regulations. It is understandable that busy practitioners could forget their ethical responsibilities in the face of these pressures, but maintaining a rigorous ethical standard is crucial to long-term success.

Written to meet the ACCE's requirements for all construction students, this textbook draws on the authors' industry experience, as well as detailed case studies to introduce and explore ethics in the construction industry. Within each chapter, the authors present the key ethical issues in important areas of construction management such as:

- Contracts and Bidding
- Documentation
- Codes and Compliance
- Discrimination and Harassment
- Client Relations.

Lists of further reading and discussion questions will help readers at all levels to develop their understanding of this issue. Written as a resource to accompany students throughout their degrees, this is the ideal book to give students or practitioners the breadth and depth of understanding required to successfully negotiate the ethical challenges facing the construction organization of today.

Rebecca Mirsky is an Associate Professor with the Department of Construction Management at Boise State University, USA. She is a registered professional environmental engineer and LEED Accredited Professional with over 20 years of experience in the construction industry, having worked for both Morrison Knudsen and the Bechtel corporations in various roles including estimator, project cost engineer, field engineer, and project engineer. She is also an active member of the American Council for Construction Education (ACCE), and participates regularly on accreditation visiting teams.

John Schaufelberger is the Dean of the College of Built Environments at the University of Washington, USA, where he has taught since 1994. A licensed professional engineer, he served thirty years as an officer in the US Army Corps of Engineers prior to joining the University of Washington faculty. He managed major public works construction projects in Asia, Europe, and the Middle East, as well as at many locations throughout the United States. He is a member of the National Academy of Construction, and author of several books on construction management. He also serves as chair of the ACCE Accreditation Committee.

Professional Ethics for the Construction Industry

Rebecca Mirsky and John
Schaufelberger

Routledge
Taylor & Francis Group

LONDON AND NEW YORK

First published 2015
by Routledge
2 Park Square, Milton Park, Abingdon, Oxon, OX14 4RN

Simultaneously published in the USA and Canada
by Routledge
711 Third Avenue, New York, NY 10017

Routledge is an imprint of the Taylor & Francis Group, an informa business

© 2015 Rebecca Mirsky and John Schaufelberger

British Library Cataloguing in Publication Data
A catalogue record for this book is available from the British Library

Library of Congress Cataloging-in-Publication Data
Mirsky, Rebecca.
Professional ethics for the construction industry / Rebecca Mirsky and John Schaufelberger.
pages cm
Includes bibliographical references and index.
1. Construction industry–Moral and ethical aspects. 2. Building–Moral and ethical aspects.
I. Schaufelberger, John, 1942- II. Title.
HD9715.A2M465 2014
174'.9624–dc23
2013048097

ISBN13: 978-0-415-67751-6 (hbk)
ISBN13: 978-0-415-67752-3 (pbk)
ISBN13: 978-0-203-80828-3 (ebk)

Typeset in Century by Fakenham Prepress Solutions, Fakenham, Norfolk NR21 8NN

Contents

Foreword

When I was approached to write a foreword for this text I quickly agreed. I see no topic more important to the future of our construction industry than honorable, professional practice. We are a service industry as well as a relationship industry and to do our work well we need the complete trust of our clients, stakeholders, and team members. When a builder takes advantage of a client he/she is doing more than damaging a relationship, he/she is also hurting the image of our industry and every builder. Warren Buffet in his book, *In The Snowball: Warren Buffet and the Business of Life*, says it well:

> Lose money for the firm, and I will be understanding. Lose a shred of reputation for the firm and I will be ruthless.

Good builders build bridges, develop trust, and look for ways to improve the image of their companies and industry. Our industry has come a long way in the improvement of our technical practices, but still needs to do a lot of work to improve our professional practice. Borrowing Warren Buffet's analogy for a moment we have a lot of work to do to push the "professional practice snowball" back up the hill since we have a long history filled with hundreds of examples of unethical practices. Unfortunately, some of our employer companies have also "institutionalized" questionable practices that confront our young graduates when they enter the workplace. To succeed our graduates need a very strong grounding in professional ethics to acquire the confidence and the knowledge to handle these pressures and additionally help change our industry. Quite a challenge!

Having taught professional ethics for over ten years, I have found that learning ethics is as much a lesson of awareness as it is practice. Practicing professionals need to be able to first identify the ethical dilemma then once the dilemma is understood he/she can figure out the best way to deal with it, oftentimes with the help of their immediate supervisors. Both the American Institute of Constructors and the Construction Management Association of America have established codes of ethics to guide the constructor. These codes are included and described in the following text.

Learning and applying good ethical practice is not easy since the breadth of ethical coverage is vast. Ethics come into play when dealing with the public:

> Your site work contractor is behind schedule and during excavation uncovers a small dinosaur bone. Do you stop the job and investigate or ignore and continue?

With subcontractors:

> You are in the process of evaluating electrical bids and your boss asks you to share the low bid to date with his preferred electrical subcontractor.

With trade workers:

> Your firm does underground utility work and your trade workers are ready to work, but the trench box which is meant to hold back the earth has not arrived. It should arrive in an hour, and the excavation is not that deep. Start work?

Other constructors:

> Your roommate, also a construction manager, works for a competing firm. Through him you have found out that they are being considered for the same project that your firm is bidding on. He has also shared some of the bid presentation material with you, which you perceive

will weigh the bid in their favor. Do you tell your own bosses about this, potentially giving your firm the edge?

Students often see these expectations as common sense until they see the situation in a case and realize how difficult it is to first separate and provide context to the ethical issue. Once the ethical issue is understood, alternatives need to be developed and the best course of action chosen. What students find painful is that often all alternatives may be bad and the best choice may very well be the decision that causes the least harm. The last example is a good instance of no good choice. Passing on the information will likely damage a personal relationship. Keeping quiet could mean losing the bid. And perhaps it still seems clear what to do until you add in the context that your firm (and your job) depends on getting this particular project whereas the other firm is in better financial health and your roommate's job secure.

I view the construction industry as an industry in transition. Traditionally, construction work was acquired through a competitive bid making the constructor responsible to complete the work as defined by the bidding documents. A constructor was obligated to complete the work safely, legally, and completely, but the builder's loyalty was to his/her company. Builders were "contractors" with all obligations defined by the contract. Times, however, have changed.

For many reasons constructors are now being hired earlier in the life cycle of a project, often on a fee basis, as owner construction consultants. In this capacity, a constructor's ethical obligations and loyalty run to the owner first. Advice provided is to be in the owner's best interest not that of the constructor. This situation gets particularly tricky when the contractor is hired early on a fee basis, but the plan is for the project to transition to an "at risk" guaranteed maximum price (GMP) contract later. Operating under the GMP, the constructor is now operating at risk with obligations defined by the GMP contract similar to the more traditional approach. In essence, pre-GMP the constructor obligations are first to the client, post-GMP they are defined by the contract.

What complicates this situation even further is that the advice the constructor provides to the client pre-GMP may later form the basis of the contract that his/her firm will operate under later when at risk. It takes a savvy builder to "navigate these ethical waters" since any misstep either real or perceived can tarnish a constructor's reputation. To be safe, some firms make the conscious decision to stick with only one mode of operation: "at risk" or "not at risk." This simplifies internal operations and the behavior of inexperienced staff that cannot possibly understand all of the ethical nuances.

In summary everyone that cares about our construction industry and its future realizes how important it is to improve its image. We need to improve safety; we need to create ways to get work done with minimal disruption to the client and public; and we need to be trusted. Many builders are trusted and are enjoying successful careers, but the general public perception of construction is that builders are not to be trusted and there are enough examples of corruption, bribery, and shoddy work to back up that perception. Realistically it takes years to restore a tarnished reputation and there are few companies that can accomplish this without consciously changing their fundamental business practices. The most effective action is to ensure that those that enter our industry have a strong ethical foundation. We can do this first in the classroom providing the knowledge, skill set, and understanding of the expected behavior and conduct of a true professional. Our professional associations through their published ethical codes and accrediting agencies through their accreditation standards insist that our new constructors are ethically trained. We are all committed to a new future and your success as a future construction industry professional.

Frederick E. Gould PE, CPC
Professor of Construction Management
Roger Williams University

1 Introduction

Learning Objectives

After reading this chapter, you should be able to:

- Describe the five common ethical approaches that form the foundation of professional business ethics.
- Explain how an ethical issue differs from a legal issue.
- Identify common areas of ethical challenges specific to the construction industry.

Ethics Fundamentals

One of the distinguishing features of human society is the desire to establish rules and order, in the form of guidelines, covenants, codes, standards, and laws that define acceptable behavior and conduct in a civilized world. Throughout history, religious leaders, philosophers, and politicians have pondered and discussed the human condition, and argued over how to live, or inspire others to live a just and moral life, whether dealing with individuals or empires. The golden rule, which requires one to treat others as one would wish to be treated, is one of the most fundamental and simple tools for making ethical decisions.

Whether or not a set of beliefs is based on religious tenets, those beliefs constitute the moral values people use to distinguish between right and wrong in making personal and business decisions. Ethics is the practice of examining those moral values in the face of day-to-day situations and having the courage to do what is right. Each decision has consequences, either to us ourselves or to others. Generally there are three primary ethical directives: *loyalty, honesty, and responsibility.* Loyalty may be requested from many groups and institutions: friends, family, employer, profession, and society. Honesty is more than truth telling. It involves not lying, but, more importantly, the correct representation of ourselves, our actions, and our views. Responsibility means anticipating the potential consequences of our actions and taking responsible measures to prevent harmful occurrences.

Whereas laws and regulations define courses of action to which we must legally adhere or avoid, ethics include standards of conduct to help us make the right decisions. For example, an individual may hold a moral belief that stealing is wrong. Aside from legal interpretations, ethics provide a framework for that individual to evaluate why stealing is wrong, or unethical, in a certain context. Perhaps it is unethical because the individual has represented himself as an honest person. Or maybe it is unethical because by stealing, the individual gives herself an unfair advantage over someone else. Or perhaps it is unethical because it deprives another of his right to the stolen item. Ethical decisions are often referred to as dilemmas because they arise from situations that are rarely black or white. We can use the reasoning behind the various ethical theories to help us to navigate the gray areas we frequently encounter in our professional and personal lives.

There are many excellent books, journals, articles, and other references that explore deeply, and in great detail, the different theories and branches of ethical thought. For our focused look at business ethics, and specifically, business ethics for construction, we will limit ourselves to five ethical approaches that we can apply to some of the most common ethical challenges in our industry.

The Utilitarian Approach – The ethical choice is whichever produces the greatest good for the greatest number of people. The Utilitarian Approach is concerned with the *consequences* of the action or decision (as opposed to the motivation or the means), and the direct or indirect impacts on everyone involved. When applying the Utilitarian Approach, it is possible that individuals may be harmed, or suffer negative consequences, as long as the majority of those affected benefit from the decision. The Utilitarian Approach also allows for using potentially unethical means such as lying to achieve the end goal of maximum benefit to the greatest number.

The Rights Approach – The ethical choice is the one that respects the fundamental rights of others to be treated as equals capable of making their own decisions. Some of the common fundamental rights are free speech, liberty, safety and security, and equal protection under the law. Rights are also related to duty; if we acknowledge that others have fundamental rights, then we have a duty to respect those rights. If we ignore that duty, then we have acted unethically.

The Justice Approach – The ethical choice is the one that treats everyone involved fairly and in accordance with what they are due. This means that people in equal circumstances should be treated (and compensated) equally. This approach is commonly applied in situations involving discrimination in the workplace. In evaluating a situation using the Justice Approach, we ask, "Are the individuals in this situation being treated equally, and if not, is there a justifiable reason for the inequality?"

The Common Good Approach – The ethical choice is the one that promotes or contributes to the common good of society or the community. This approach is based on the assumption that there are certain services and resources, either natural or human made, whose existence benefits our common good as a society, and that the ethical choice preserves and protects those resources. Examples of services and resources that we all benefit from might include a healthy environment, a robust healthcare system, high quality education, and modern infrastructure. For all of society to benefit from resources that support the common good, all of society must also contribute to and share in the protection of those resources. Problems and conflicts arise when people value resources differently, choose to use more than their fair share, or refuse to conserve so that others may also benefit.

The Virtue Approach – The ethical choice is the one that arises from and demonstrates our moral virtues. Honesty, integrity, respect, trust, and fairness are all examples of virtues. Companies commonly include the virtues they value most highly within their stated core values. These are understood to form the foundation for the corporate culture and how the company conducts its business. When faced with an ethical dilemma, the Virtue Approach requires that the individual or the company act in such a way that personifies the virtues they want to be known for.

Applying an ethical approach in a business setting means using our knowledge of ethical decision-making in the management and operations of that business. Because there can be pitfalls with any single ethical approach when used alone, a sound framework for ethical decision-making looks at a situation from the perspective of each approach. Chapter 6 has more information about using a framework to make ethical decisions.

Ethical Challenges in the Construction Industry

Human survival depends on the protection and services provided by our built environment. Laws, codes, and standards to ensure safety and quality have been an integral part of the construction industry for hundreds of years. The earliest known building code is found in the Code of Hammurabi, dating back to nearly 1,800 BC. The Babylonian king Hammurabi believed in strict punishment for shoddy work, stating:

> If a builder has built a house for a man, and has not made his work sound, and the house he built has fallen, and caused the death of its owner, that builder shall be put to death.[1]

Today we address building safety and quality through building codes, safety regulations, and a wealth of specifications and standards. But many of the decisions about how a construction company conducts business are not so well defined. A company's success is tied to its reputation, and that reputation is built on relationships with clients, partners, subcontractors, suppliers, employees, agencies, and the community.

In construction, we face many ethical dilemmas in contract procurement, cost estimating, project management, accounting/financial management, customer relations, subcontractor relations, and material vendor relations. A construction company's ability to acquire and retain customers as well as to attract and retain talented employees is directly influenced by its reputation for ethical culture and the ethical behavior of its leaders and employees. Construction is basically a service industry, and purchasers of construction services have choices. Thus a reputation for ethical behavior is essential for success. A company's ethics needs to be one of its core values to guide its employees in making daily decisions. Mishandling ethical situations can damage the company's reputation and destroy the morale of its employees.

The construction industry is also extremely competitive with ample temptation to look for advantages by cutting corners, making backroom deals, favoring friends or relatives, or withholding information. At times, it can seem that the ethical decision runs contrary to the financial success of the company, and it can be extremely difficult for companies and their employees to know how to proceed. This is particularly challenging in the management of construction projects which rely on successful collaboration of individuals representing diverse organizations with different ethical cultures. Treating everyone fairly should be emphasized in the execution of construction projects.

In the following chapters, we look at some of the most common ethical challenges encountered by construction professionals, and provide a framework for making tough decisions. The examples and scenarios are taken from our own real life experiences as well as those of our colleagues in industry. We look into some of the dilemmas associated with obtaining and performing the actual work of construction, including issues arising from relationships with clients, subcontractors, and suppliers. We also discuss issues with discrimination and harassment in the office and on the jobsite, and examine current practices in creating and implementing company codes of ethical conduct. Finally we address some emerging topics that are presenting unique ethical challenges, such as the increase in digital information, prevalence of social media, and the growing interest among corporations in social responsibility.

References

For more detailed information and an excellent discussion about the five ethical approaches presented in this introduction, please visit the material on ethical decision-making, available on the website of the Markkula Center for Applied Ethics at Santa Clara University, at http://www.scu.edu/ethics/.

Note

1 Hammurabi's Code of Laws, #229.

2   Documents and Bidding

Learning Objectives

After reading this chapter, you should be able to:

- Describe the ethical challenges that can occur during the process for submitting a bid for a project between:
  - a contractor and a project owner
  - a contractor and a project designer
  - a contractor and subcontractors.
- Describe the ethical challenges that can occur between a contractor and suppliers during the bidding process.
- Imagine and evaluate potential consequences resulting from ethical decisions made during the contract procurement process.

Introduction

In this chapter, we will examine some of the ethical situations that one may encounter in the procurement of a construction contract or in the development of a bid or proposal for a construction project in response to a project owner's solicitation. Documents and bidding ethics is about the basic concepts and fundamental principles of decent business conduct on or before the submission of a bid or proposal.

Competitive bidding is one form of contract procurement that a project owner may use to select a general contractor for a construction project. In this process, a project is described in the bid documents, and prospective contractors are requested to submit bids or prices to construct the described project. Since price typically is the criterion used for award of bid contracts, the bidding process is seen as a "market driven" process in which the lowest bid represents the "best value."

During the bidding process, the project owner and project designer usually conduct a pre-bid meeting with prospective bidders and subcontractors to address any issues that they have identified as a consequence of reviewing the contract documents. At the conclusion of the pre-bid meeting, the project owner collects the issues identified by prospective bidders and subcontractors and provides responses to all prospective bidders in the form of a contract addendum. This use of contract addenda ensures that all prospective bidders are using the same project information when developing their bids and ensures fairness and equitable treatment of all prospective bidders during the bidding process.

Project owners may choose to select a general contractor for a project by requiring prospective general contractors to submit competitive bids or to submit proposals for a negotiated selection process. Competitive bids may be submitted on a lump-sum or unit-price basis or a combination of both. Negotiated proposals may use the same methods of pricing, or often may use a cost-plus approach in which most direct project costs are reimbursable and other contractor costs are included in the fee. Ethical issues can occur during both procurement processes as we will discuss in this chapter.

When developing bids or cost proposals to submit to project owners, the general contractors decide which scopes of work they will perform with their own work forces and which scopes of work will be subcontracted to specialty contractors. The subcontracted scopes of work are organized into subcontract bid packages, and prospective subcontractors are invited to submit quotations for each subcontract bid package. The general contractors evaluate the subcontractor quotations and decide which ones to select as a part of their bid or proposal preparation process. Subcontracts are not awarded, however, until the general contractor receives the contract from the project owner.

There are many legal issues associated with the contract procurement process, such as the contractors meeting to discuss their bids and deciding which one would submit the lowest bid or contractors offering bribes to be selected. None of these legal issues are addressed in this chapter. We will restrict our discussion solely to ethical issues that may occur during the contract procurement process.

Introductory Case Study

A project manager for Acme Construction reviewed the construction drawings and specifications for the construction of a shopping center prior to attending a pre-bid job site tour. During the review, the project manager identified two errors in the elevations provided in the drawings. During the site visit, the project manager asked the project architect about the elevation errors and was provided the correct information. However, the project owner did not issue an addendum to all prospective bidders making the elevation corrections to the project drawings. Were the project owner's actions ethical?

The site work associated with the project was unit priced, because a portion of the site contained contaminated soil that needed to be removed and replaced, additional fill material needed to be imported, and a large asphalt parking lot constructed. In addition, major utilities were to be installed on the site. During the review of the contract drawings, Acme Construction's estimator determined that the quantity shown on the unit price bid sheet for asphalt pavement was considerably less than what would be required for completing the project. The estimator decided not to notify the project owner and to inflate the unit price for the asphalt bid item because of the anticipated overrun. Was the estimator's action ethical?

Quotations were solicited from six prospective electrical subcontractors for the project. The lowest quotation was submitted by Northern Lights Electrical Contractors, but the project manager preferred to work with West Coast Electric. The project manager contacted the owner of West Coast Electric and provided the quotation received from Northern Lights and told West Coast that they could have the job if they revised their quotation to a value less than that submitted by Northern Lights. Was the project manager's action ethical?

The shopping center structure was to be constructed of steel. Acme's estimator solicited quotations from three steel suppliers for the project. The estimator was concerned both about the cost of the steel and the ability of the suppliers to meet the required delivery dates established in the preliminary construction schedule. Continental Steel submitted the lowest quote but did not guarantee that they could meet the required delivery dates. The salesman for the steel supplier indicated to Acme's estimator that if Continental Steel received the supply contract they would host the estimator to a fishing trip. What should the estimator do in this situation?

Ethical Challenges

Ethical Challenge: Errors in Project Documents

The bidding instructions given to prospective bidders on a project typically require that the bidder consider all conditions described in the contract documents and all conditions that can be observed by physically visiting the site. Liability for hidden conditions not described in the documents or in a soils report typically is the responsibility of the project owner. These would include buried utility lines not shown on the drawings or contaminated soil not described in the documents.

During a pre-bid conference on the job site, representatives of the project designer and the project owner are present to describe the project and collect inquiries from prospective general contractors and subcontractors regarding the contract documents. To ensure that everyone who participates in the bidding process has the same information, the project owner should collect all of the questions and issue a contract addendum providing appropriate responses to each question. From the perspective of the justice approach, it is unethical to provide answers only to the party who asked the questions. Even though the issuance of a contract addendum late in the bidding process may necessitate delaying the receipts of bids, it is the ethical responsibility of the project owner to do so. It may also negate the need to issue a change order after the contract has been awarded.

Ethical Challenge: Bid Shopping

Bid shopping occurs when general contractors disclose to prospective subcontractors the price quotations received from competing subcontractors. The intent is to encourage subcontractors to lower their prices. Again, based on the justice approach, this is considered unethical because it discloses information that is confidential, and not available equally to all bidders. A likely result is subcontractors refusing to work with general contractors who use this practice. The subcontractors are being asked to provide their best price for a specific scope of work, and they provide the price to the general contractor with the expectation that their price will not be shared among their competitors. Often subcontractors' quotations contain lists of specific inclusions and specific exclusions, which means that the scope of work addressed by each subcontractor may vary. This requires the general contractor to carefully evaluate each quotation and select the ones that provide the best value to the general contractor.

Another form of bid shopping that is unethical is when a general contractor uses the quotation of one subcontractor in their bid, but selects a different subcontractor to perform the work. For example, suppose Allied Construction Company is developing a bid for the construction of a high school and solicits quotations for the electrical work associated with the project. Capital Electric submits the lowest quotation for the electrical work, and their price is used by Allied in preparing their bid to submit to the project owner. Allied receives the contract for construction of the high school, but instead of awarding the subcontract for the electrical work to Capital Electric, they contact Southwest Electric and offer them the subcontract if they will do the work for less than the price submitted by Capital Electric. This sharing of Capital Electric's proposed price with another subcontractor is considered a form of bid shopping and is unethical. It is also dishonest, and therefore violates the approach of virtue-based ethics.

Ethical Challenge: Receipt of Favors

Subcontractors may offer favors to general contractors in an effort to win a subcontract, and suppliers may offer favors to secure a contractor's business. Such practices may be unethical. Sometimes suppliers offer their good customers discounts for early payment of their invoices, and such practices are not considered unethical. However, a supplier offering a personal favor to the contractor's employees would be considered unethical. Whether or not a person's behavior is influenced by the receipt of a favor, there is a perception that such actions may occur. Anyone involved in making decisions related to award of contracts or subcontracts needs to ensure that a no-favor policy is adopted. This may include tickets to athletic events, fishing trips, meals, or other social events. It is best to not enter into a situation where there is a perception of favoritism or unfair advantage.

Applicable Standards

The applicable standards are to practice good faith and fair dealing in the solicitation of bids or proposals and in the preparation of bids or proposals for construction projects. Project owners and designers need to ensure that all prospective bidders have the same information relative to project scope and conditions. Any issues identified by the prospective bidders during the bidding process need to be resolved, and the information provided to all bidders. Contractors often are more willing to bring issues identified in the contract comments to the attention of project owners when a negotiated process is to be used to select the contractor to whom to award the contract. This is because the criteria to be used to select the contract recipient may include several factors in addition to price. In a bid process, the proposed contract price typically is the primary criterion for selecting the general contractor to whom to award the contract.

General contractors need to treat subcontractors' price proposals as confidential and not disclose the information to other subcontractors. General contractors may wish to engage in discussions with a subcontractor regarding their inclusions, exclusions, and price but should not disclose the proposals submitted by other subcontractors.

Anyone engaged in a contract procurement action needs to avoid the perception of favoritism. This is true whether the individual is employed by the project owner, the designer, the general contractor, a subcontractor, or a supplier. Acceptance of favors undermines the perceived fairness of the procurement action.

Construction Participant Perspectives

To minimize the potential for bid shopping, many government agencies and other project owners require that a listing of major subcontractors to be used on a project be submitted along with completed bid documents. This does not reduce the potential for bid shopping prior to submission of the bids, but it does reduce the potential for bid shopping after the contract has been awarded.

General contractors who conduct bid shopping may find that quality subcontractors are not interested in working for them and may not respond to the general contractor's solicitation for subcontract quotations. This is particularly true during periods when the subcontractors have adequate backlogs of work. Specialty contractors want general contractors to treat the quotations that they submit as confidential information that is not shared with their competitors. How the subcontractors are treated will greatly influence their interest in working with a general contractor and the price that they request for their services.

Subcontractors estimate their cost of doing business with each general contractor and may propose different prices for similar bid packages from different general contractors. This practice is not considered unethical and is the reason why general contractors should treat subcontractors fairly. When a subcontractor has a good experience working with a general contractor, they often propose very competitive prices for subcontract scopes of work on future projects.

Some project owners may choose to select general contractors for their projects using a negotiated procedure. In this process, the relationship between the owner and contractor is more collaborative than it is in a bid procedure. The proposed construction schedule, experience of the contractor's project team, the contractor's reputation for quality work, the contractor's safety record, and other factors may be used in addition to project cost in the selection process.

Questions and Scenarios for Discussion

Evaluate the following scenarios.

1 Near the end of the bidding process for the construction of a hotel, a project owner received a written inquiry from a prospective bidder regarding some missing information on the structural drawings. The owner consulted the structural engineer for the project to determine the requested information and provided the information to the prospective bidder. Not wishing to delay receipt of bids for the project, the project owner decided not to issue an addendum to the contract documents.

- Were the actions of the project owner ethical?
- How would you have handled this situation?

2 Continental Constructors received a contract for the construction of a hospital. During the bidding process, Continental received quotations for the mechanical scope of work from six subcontractors. Five of the quotations were solicited, and the sixth was unsolicited. Continental selected the lowest quotation from among the five solicited quotations, because they had had a previous unsatisfactory experience with the subcontractor who submitted the unsolicited quotation. However, the unsolicited quotation was the lowest price.

- Was it unethical for Continental Constructors to not select the lowest price quotation?
- What would you have done in this situation?

3 You are finalizing the bid for the construction of a middle school, and you notice an omission in the bid for exterior painting. You insert an

estimated amount to cover scope of work. Your company is successful in receiving the contract, and you contact a painting subcontractor and tell them that they can have the subcontract for the painting work if their price does not exceed the amount that you estimated.

- Are you treating the painting subcontractor ethically?
- What would be an alternative approach to solving the problem?

4 A general contractor is developing a proposal for the construction of a medical clinic. Since the contract will be executed prior to the completion of design, the contract will be negotiated on a cost-plus-fixed-fee basis with a guaranteed maximum price. There is significant site work required for the project, and the contractor has decided to ask five subcontractors for quotations for the site work bid package. The contractor's estimator received the five quotations and upon review determined that the preferred subcontractor did not submit the lowest price. The estimator then contacted the preferred site work subcontractor, told them that their price was 10 percent higher than the lowest quotation received, and asked the subcontractor if they would like to lower their price by 10 percent in order to receive the subcontract.

- Were the actions of the general contractor's estimator ethical?
- If you were the subcontractor who was contacted, what would be your reaction?

5 You are the estimator for Excel Mechanical Contractors and have received requests for quotation from five general contractors for the mechanical scope of work associated with the construction of a research facility. You evaluate the scope of work and your past experiences in working with each of the general contractors. You choose to submit different prices on each of the quotations provided to the general contractors. The reason for the different prices was your perceptions regarding how you would be treated by each of the general contractors.

- Was submitting different prices to each of the general contractors ethical? Why or why not?

3 Construction Contracts and Purchase Agreements

Chapter Outline

- Learning Objectives
- Introduction
- Introductory Case Study
- Ethical Challenges
- Applicable Standards
- Construction Participant Perspectives
- Questions and Scenarios for Discussion

Learning Objectives

After reading this chapter, you should be able to:

- Describe the ethical challenges that can occur during the administration of a construction contract between:
  - a contractor and a project owner.
  - a general contractor and a subcontractor.
- Describe the ethical challenges that can occur between a contractor and a supplier during the administration of a purchase agreement.
- Imagine and evaluate potential consequences resulting from ethical decisions made during the administration of contracts and purchase agreements.

Introduction

When a project owner and a contractor sign a construction contract, the responsibilities of each party are limited to those prescribed in the contract documents. The project owner agrees to provide information and to compensate the contractor for the scope of work described in the contract. The contractor agrees to complete the project described in the drawings and specifications within the specified duration. Few general contractors actually perform all of the work required to complete the project, but instead, select specialty contractors to perform selected scopes of work as subcontractors under the terms and conditions of their subcontracts.

Both general contractors and subcontractors purchase construction materials from suppliers by means of purchase orders or agreements. The contractors and subcontractors provide the contract specifications to the supplier to enable the supplier to determine which products meet the specification requirements and provide the information to the general contractors and subcontractors. If the contract stipulates that the material must be approved by the project designer, the contractor must provide a submittal that either contains a sample of the material or manufacturer's information demonstrating conformance with the contract specifications. Once the material has been approved by the project designer, the contractor or subcontractor determines the quantity required and issues a purchase order to the supplier requesting the cost for the approved materials. The contractor or subcontractor then approves the purchase agreement ordering the required materials. A purchase agreement basically is a supply contract for the acquisition of construction materials.

There are many legal issues associated with administration of construction contracts and purchase agreements. None of these legal issues are addressed in this chapter. We will restrict our discussion to ethical issues that may occur during the administration of these contracts.

Introductory Case Study

Continental Construction Company has been awarded a contract for the construction of a medical office building. During the development of a bid for the project, the construction company's project manager decided to hire a subcontractor for the electrical scope of work. During the bidding process, the company estimator solicited quotations from five electrical contractors. The best value quotation was received from Capital Electric, and their quotation was used by Continental's estimator in developing the winning bid for the project. Continental's project manager had good experience with Olympic Electric and would like to use them on this project. Olympic had submitted a quote for the electrical work, but it had been 10 percent higher than the price submitted by Capital. The project manager contacted the owner of Olympic Electric and told him that Olympic would be awarded the electrical subcontract if they agreed to perform the work at the price quoted by Capital Electric. Were the actions of Continental's project manager ethical?

The contract that Continental signed for the construction of the office building allowed the submission of materials that deviated from the specifications provided that the contractor indicated such deviations on the submittal documents and provided a justification for proposing a deviation. The mechanical subcontractor on the project proposed to use a fan unit that did not comply with the contract specifications. The subcontractor did not indicate on the submittal documentation that the proposed fan unit was a deviation from the contract specifications. Upon reviewing the subcontractor's submittal, Continental's project engineer recognized that the proposed fan unit did not conform to contract requirements, but chose to forward the submittal to the project designer without comment. Were the actions of Continental's project engineer ethical?

The project owner decided to make some changes in the layout of the building lobby and asked Continental's project manager for a cost proposal for the additional work. The project manager reviewed the additional scope of work and estimated that the direct cost for the work would be $500,000. Project indirect cost, company overhead, and profit would add $100,000 because executing the change order would delay the project completion by one week. Work on the project was ten days behind schedule, and the project manager decided to include the overhead cost and potential liquidated damages liability to the change order cost, which he estimated to be $150,000. Thus the cost that the project manager submitted to the project owner for the change order to the additional lobby work was $750,000. Were the project manager's actions ethical?

Construction has been underway on the project for five months, and Continental Construction has been experiencing cash flow problems on the project. The company's project manager decided to mitigate the cash flow problem by not paying subcontractors for the work that they had performed until 45 days after the general contractor received payment from the project owner for the work. In effect, the project manager was using the money owed to the subcontractors to fund the general contractor's cash flow issue. Was the project manager's action ethical?

Ethical Challenges

Ethical Challenge: Award of Subcontracts

In developing a bid or proposal for a construction contract, the general contractor's estimator typically invites multiple specialty contractors to submit quotations for each subcontract work package. These quotations are evaluated, and the best value subcontractor is selected for each subcontract work package. If the general contractor is selected for receipt of the construction contract, the specialty contractors submitting the best value quotations are offered subcontracts for the project. From the perspective of the justice approach, contacting other specialty contractors and telling them the quotations provided by their competitors is a form of bid shopping and considered unethical.

Ethical Challenge: Errors in Submittal Documentation

When engineers and architects prepare their designs for a construction project, they rarely specify single products or materials to be used on the project. They typically prepare specifications that describe the types of materials or products that are acceptable. To ensure that the materials and products selected by the contractors conform to the design intent, the construction contract requires the submission of product samples or manufacturers' literature describing the proposed materials or products. These are known as submittals. The submittals must be approved by the designer before the contractors issue purchase orders for the acquisition of the proposed materials or products. To avoid delay of the project, the general contractor's project team needs to ensure that all proposed materials and products fully conform to contract requirements. If a deviation is proposed, it must be highlighted in the submittal documentation. If deviations are not noted, the submittal may be disapproved or unacceptable materials may be purchased for the project.

Ethical Challenge: Pricing of Change Orders

During the construction of a project, the project owner may wish to change some aspect of the project or there may be some discrepancy in the contract documents. In either case, the result would be a change to the contract, known as a change order. The project designer typically prepares the change order, and the general contractor is asked to submit a cost for executing the additional work. Estimating the direct cost of the work usually is not difficult. Estimating the impact cost on the planned execution of the project, however, may be more difficult. The contractor's project manager needs to ensure that the proposed change order cost includes only the direct cost of doing the work, the indirect cost, profit, and any impact cost of doing work out of sequence. It is dishonest, and therefore unethical based on the virtue approach, to include costs that are not the result of the change order.

Ethical Challenge: False Claims

Sometimes a contractor encounters conditions while executing a construction project that were not depicted in the contract documents. The contractor typically places the project owner on notice and requests a change order to cover the additional costs incurred. If the project owner declines to issue a change order, the contractor may file a claim for additional compensation. Sometimes contractors inflate their costs when submitting their claims. Submitting a false claim is an illegal action under the False Claims Act if the project owner is part of the federal government (Chapter 8 Client Relations has more information about the False Claims

Act). Submitting a false claim is unethical if the project owner is a private entity or state or local government.

Ethical Challenge: Payment to Subcontractors

Because many general contractors subcontract major portions of a construction project, the work of the subcontractors is critical to ensure a successful project. When working for a general contractor, the subcontractors expect to be treated fairly. This means ensuring that the job site is ready for each subcontractor when they are scheduled to arrive on site and ensuring that they are paid in a timely manner. The subcontractors must pay their employees prior to receipt of payment from the general contractor, but usually pay for their construction materials after receipt of the general contractor's payment. The subcontractors' requests for payment are included in the general contractor's monthly request for payment to the project owner. Once the general contractor receives payment from the project owner, the subcontractors need to be paid. The subcontract document may provide for some retention, perhaps 5 percent, from the payment to the subcontractor, but the remainder needs to be paid promptly to enable the subcontractors to manage their cash flow.

Ethical Challenge: Material Purchasing

Just as the situation with subcontractors, suppliers should be treated in an ethical manner. A general contractor may solicit quotations from three concrete suppliers for concrete to be delivered to a project site. Each quotation should be treated as proprietary information. From the perspective of the justice approach, it is not ethical to provide competitors' cost proposals in an effort to obtain a better price from a preferred provider. Suppliers also need to be paid in a timely manner, just like subcontractors. Some suppliers may offer discounts for early payment, but many purchase orders state that the contractor will pay for the materials once payment has been received from the project owner.

Applicable Standards

The applicable standards are to practice good faith and fair dealing in the administration of construction contracts and purchase agreements. General contractors need to be fair in the selection of subcontractors for their projects and not engage in bid shopping. Subcontractors and suppliers need to be treated fairly during project construction and paid in a timely manner. Subcontractors' and suppliers' willingness to work with a general contractor is greatly influenced by their perspectives of how they will be treated by that contractor.

Any documents submitted to general contractors or submitted to project owners by general contractors must be completed accurately and contain all of the information required by the contract. Not providing complete information may delay the processing of the documents or may result in improper decisions being made. Change orders and claims need to be correct and not include requests for reimbursement for unrelated issues.

Construction Participant Perspectives

When project owners select general contractors for construction projects, they expect the contractors to deliver quality projects that conform to the contract requirements by the dates specified in the contracts. To be able to meet the project owners' expectations, general contractors rely on subcontractors to perform selected portions of the projects and suppliers to provide the needed construction materials. To ensure that quality subcontractors and suppliers are willing to work with a general contractor, the contractor needs to treat them fairly.

When subcontractors and suppliers submit quotations for their scopes of work or for construction materials, they expect the general contractor to treat the information as proprietary and not disclose the information to their competitors. How the subcontractors and suppliers are treated greatly influences their willingness to work with a general contractor. Subcontractors who have had a bad experience with a general contractor may choose not to provide a quotation for one of the contractor's projects when invited to do so.

Project owners expect general contractors to comply fully with all contract requirements. The contract specifications identify what material submittals are required for the project and contain descriptions of what materials are acceptable for use on the project. If deviations are to be proposed, they must be clearly identified in the submittal documentation. Failure to identify deviations violates the terms of the contract, but it also is unethical. It may result in the contractor purchasing and installing improper materials that later will need to be replaced at the contractor's cost.

Project owners expect that change orders and claims are priced fairly. They realize that changes in the scope of work often have cost impacts and may have schedule impacts. Contractors are expected to assess the impact of the change order on the planned construction of the project and only include costs associated with the change. Adding unrelated costs is not ethical and may result in a dispute regarding the cost of the change order.

Questions and Scenarios for Discussion

Evaluate the following scenarios.

1   A contractor has a contract for the construction of an office building for an agency of the federal government. One of the contract specifications requires that all materials used on the project be manufactured in the United States. A steel supplier has offered steel for the project at a very competitive price. The general contractor's project manager accepts the proposal and requests submittal documents. The steel supplier provides the required documents but omits a certificate of origin for the steel. The general contractor's project engineer reviews the submittal, contacts the supplier, and determines that the steel was manufactured in Mexico. Because this supplier's price was lowest for the steel needed for the project, it was used in developing the project budget. Not wanting to exceed the budgeted amount for steel, the project engineer chose to forward the steel supplier's submittal to the structural engineer with no indication of the place of manufacture.

- Were the actions of the project engineer ethical?

- How would you have handled the situation?

2 Eastern Construction has a contract for the construction of a data center. During the construction of the building, the contractor found that the electrical and mechanical drawings contained several conflicts with the structural drawings. Eastern's project manager submitted several Requests for Information (RFI) to determine how to resolve the conflicts. Upon analyzing the responses to each RFI, the project manager determined that Eastern would incur additional costs as a consequence of the additional work to be performed. To seek compensation for the additional work, the project manager submitted a request for a change order to the project owner, which was refused. Since the contractor had incurred unanticipated cost due to the design conflicts, the project manager submitted a claim to the project owner in accordance with the claims procedures contained in the construction contract. Eastern's electrical subcontractor had incurred additional cost due to unanticipated material price escalation for electrical materials. The project manager decided to include that additional cost in the claim for additional compensation for the design conflicts.

- Were the actions of the project manager ethical?
- How would you have handled the situation?

3 During the excavation for the foundation for an office building, Western Construction encountered a buried fuel tank and a natural gas line that were not shown on the contract drawings. The contractor's project manager placed the project owner on notice of the unanticipated fuel tank and gas line. Because of extra work associated with the fuel tank and gas line, Western's project manager requested a change order. The project owner agreed and requested a cost for the extra work associated with removal of the fuel tank and the gas line. The project manager knew that a mistake had been made in estimating the amount of concrete for the foundation. She decided to add the cost of the additional concrete to the cost of removing the fuel tank and gas line and proposed a price of $100,000 for the change order.

- Were the actions of the project manager ethical?
- How would you have handled the situation?

4 Acme Construction has a contract for the construction of a medical research building. The project requires significant site work which has been subcontracted to an excavation subcontractor. The subcontract agreement between Acme and the excavation subcontractor contains no provisions for withholding payment for work completed, but does state that Acme will make payment to the subcontractor once payment for the work has been received from the project owner. Acme's project manager has evaluated the contractor's cash flow on the project and determined that it will be negative at the beginning of the project. To reduce the amount of negative cash flow, the project manager decided to withhold payment to the subcontractor until 60 days after receipt of the first payment from the project owner and 30 days after the second payment is received.

- Were the actions of the project manager ethical?
- How would you have handled the situation?

5 Midwest Construction has a contract for the construction of a three-span highway bridge. The deck stringers of the bridge are to be precast concrete t-beams that are to be post-tensioned. The contractor's project manager contacted three precast concrete vendors to obtain quotations for the concrete girders. On a previous project, the project manager had good experience with Quality Precast, but they submitted the highest price for the concrete girders. The project manager called the owner of Quality Precast and told her that they would receive the purchase order if they lowered their price to equal the lowest quotation received.

- Were the actions of the project manager ethical?
- How would you have handled the situation?

4 Documentation and Recordkeeping

Chapter Outline

- Learning Objectives
- Introduction
- Introductory Case Study
- Ethical Challenges
- Applicable Standards
- Construction Participant Perspectives
- Questions and Scenarios for Discussion

Learning Objectives

After reading this chapter, you should be able to:

- Describe the ethical challenges that can occur related to the preparation and submission of documents used in the management of construction projects.
- Imagine and evaluate potential consequences resulting from ethical decisions made in the preparation and submission of documents and records.

Introduction

Documentation and recordkeeping are an essential part of the management of construction contracts. This is true for paper documents as well as for electronic documents. It is very important that documents be prepared properly and accurately and retained in locations for easy retrieval. There are many documents that must be retained on a construction project. Those discussed in this chapter include:

- payroll data showing hours worked and wages paid;
- safety records including descriptions of any accidents;
- reports of any environmental incidents;
- project labor cost data; and
- requests for reimbursement of expenses paid.

Some of these documents must be submitted to project owners or public agencies, while others are used by the contractors in managing their companies. The proper way to handle these documents ethically is to submit all required documentation and to ensure that each document accurately depicts occurrences during the completion of the project.

Introductory Case Study

Mountain Construction Company has a contract for the construction of five miles of highway including the replacement of a three-span bridge. Since the project owner is a state department of transportation, the construction contract contains minimum wages that must be paid to workers employed on the project site. These are known as prevailing wages and vary by trade classification, such as carpenter, equipment operator, or steel worker. The required minimum wages for each trade classification are contained in the contract specifications. Mountain Construction hired a drilling contractor to drill the holes for the construction of drilled pier foundations for the bridge. As required by its subcontract, the drilling subcontractor submitted their payroll data each month. Mountain's project manager reviewed the payroll data and observed that several subcontractor employees were misclassified and were paid lower wages than they would have received if properly classified. The project manager decided not to challenge the drilling subcontractor and forwarded the payroll data to the project owner without comment. Were the actions of the project manager ethical?

During the construction of the concrete forms for construction of the bridge abutments, two carpenters employed by Mountain were slightly injured and were unable to work for the remainder of the day. Not wanting to adversely impact the company's safety record, Mountain's superintendent chose not to report the accidents. Were the actions of the superintendent ethical?

During the fueling of a tractor on the job site, a worker spilled diesel fuel on the ground. Because it was raining, the rain washed some of the spilled diesel fuel into an adjacent stream. The equipment foreman found the spill and directed that it be cleaned up to include removal and proper disposal of the contaminated soil. Because all evidence of the spill had been removed, the foreman chose not to report the fuel spill. Were the actions of the foreman ethical?

When reviewing the cost reports for the construction of the foundation of the bridge, Mountain's superintendent noticed that the concrete work package was over budget, while the work package for the embankment construction was under budget. Not wanting to have any work packages exceed their budget amounts, he decided to transfer some of the labor charges from the concrete work package to the embankment work package. The end result was that both work packages were now within their budgeted amounts. Were the actions of the superintendent ethical?

Mountain's project engineer had been transporting concrete cylinders to a testing laboratory as part of the company's quality control process. When submitting his request for reimbursement of travel expenses, he included a parking receipt for a movie that he attended the previous evening. Were the actions of the project engineer ethical?

Ethical Challenges

Ethical Challenge: Submission of Payroll Data on Government Projects

The federal government and most state and local governments require the payment of prevailing wages to all workers employed on their projects of value greater than some specified value. The prevailing wage rates for federal projects are determined by the U.S. Department of Labor, and the prevailing wage rates for state and local government projects typically are determined by a state government agency. On prevailing wage projects, contractors and subcontractors generally are required to submit certified payroll data showing that all workers employed on the project were paid at least the prevailing wage rate. Workers may be paid higher rates, but not lower rates. Failure to pay at least the prevailing wage rates can subject the employer to steep fines, and falsification of payroll data can subject the employer to legal action. Sometimes contractors and subcontractors improperly classify workers in order to pay them lower wage rates. Such action is considered unethical and illegal. Calling workers individual contractors to avoid payment of prevailing wages is also unethical and illegal.

Ethical Challenge: Recording and Submitting Accident Data

Construction contractors are required to collect exposure and accident data for periodic reporting to a state agency or to the Occupational Safety and Health Administration (OSHA). Accident data is used by the agency to develop an experience modification ratio, which is used to determine the company's cost for workers' liability insurance coverage of its employees. Sometimes contractors fail to report minor accidents even though they meet the lost time criteria for reporting. It is unethical to fail to report accident data, and contractors found to be submitting incomplete accident data may be subjected to fines or legal action.

Ethical Challenge: Proper Environmental Records

Environmental regulations and laws generally require contractors to report any environmental incident, such as a fuel spill or discharge of untreated storm water, and require proper documentation and disposal of any contaminated materials removed from a construction site. Failure to comply with these regulations and laws may result in significant fines or legal action, but they are also considered as unethical behavior. Sometimes special environmental permits are required in order to complete a project, and the contractor must comply with the terms and conditions of the permits.

Ethical Challenge: Time Sheets

The proper reporting of employees' time on a construction project can present an ethical challenge. The correct procedure is to charge the time to the appropriate cost code for the work being performed. Sometimes a work item may be over budget, and there is pressure not to exceed the project budget. This may lead to consideration of charging employees' time to work items that are under budget. This invalidates the project cost reports by providing incorrect information resulting in an inability to compare actual costs to projected or estimated costs. Improper charging of employees' time is considered unethical. In the long run, misreporting time charges also hinders a contractor's ability to accurately estimate future projects based on historical data.

Ethical Challenge: Expense Reports

Sometimes a contractor's employees incur costs associated with the completion of a construction contract and request reimbursement from the company. Typically receipts are required in order to request reimbursement. It is unethical to request payment for expenses that either were not incurred or were for personal gain and not related to the project.

Applicable Standards

The applicable standards are to ensure that all people working on a project are treated fairly and that all documentation is accurately and completely prepared. Proper classification of workers and payment of prevailing wages are required on projects executed for government owners. Any environmental incidents and accidents must be reported as required by regulation. Labor charges should be applied to the appropriate work items so that construction company project managers can determine the causes of cost variances at the conclusion of the project.

Construction Participant Perspectives

Compliance with prevailing wage requirements on public projects is a legal requirement. This includes ensuring that all workers employed on the projects are classified properly. General contractors observing non-compliance by subcontractors must take action to bring them in to compliance.

Submission of accurate accident data is required by regulations. All reportable accidents must be reported, and the causes of accidents determined. A company's safety record is an important factor in selecting contractors for projects, but failure to report accidents can lead to significant fines and legal costs.

Contractors must comply with environmental regulations regarding the reporting of environmental incidents and control of any materials leaving the project sites. Failure to comply with these requirements can lead to significant fines and potential legal fees.

Proper accounting of labor hours is not a legal requirement, but it is an ethical issue. Company leaders need accurate information regarding the cost incurred on projects and how the actual costs compare with the estimated costs. This means that time sheets need to be properly completed, and the labor costs charged to the correct work items.

Questions and Scenarios for Discussion

Evaluate the following scenarios.

1   Pacific Constructors has a contract for the construction of an office building for a state government agency. The contract specifies that prevailing wages must be paid to all workers employed on the project site and lists the minimum wages to be paid to each trade classification. As required by their subcontracts, all subcontractors are required to submit payroll reports to Pacific each month. The reports are combined with a payroll report for Pacific's workers, and the entire report is submitted to the project owner. Pacific's project engineer reviewed the payroll reports submitted by the mechanical and electrical subcontractors. He observed that several journeyman electricians and plumbers were listed as helpers, and thus paid lower wages than journeymen would have earned. Rather than have a conflict with the two subcontractors, the project engineer decided to forward their payroll reports as submitted.

- Were the actions of the project engineer ethical?
- How would you have handled the situation?

2   Eastern Construction has a contract for the construction of a medical clinic. The contractor's project engineer went to a local material supplier to purchase several items for the project. He charged the needed items to a company credit card used for purchasing project materials. During the trip, the project engineer stopped at a restaurant and charged his lunch to the same credit card.

- Were the actions of the project engineer ethical?
- How would you have handled the situation?

3   Western Construction has a contract for the construction of a building that will house an automobile dealership. During the construction, the site work contractor encounters contaminated soil. The amount of soil appears to be minimal, so the subcontractor decides to reuse the contaminated soil as fill under the dealership parking lot.

- Were the actions of the subcontractor ethical?
- How would you have handled the situation?

4   Acme Construction has a contract for the construction of a water treatment plant. There is significant site work needed to prepare the site for construction of the plant. Both the electrical and mechanical work will be performed by subcontractors, but Acme will self-perform most of the remaining work. As the project progresses, the superintendent notices that the work package for the site work is over budget because of lower than anticipated worker productivity. To minimize the effect on the project budget, he decided to charge some site work labor hours to other work packages that were completed under budget.

- Were the actions of the superintendent ethical?
- How would you have handled the situation?

5 A contractor has a contract for the construction of a warehouse building that will be constructed with tilt-up concrete wall panels. During the construction of the wall panels, a concrete finisher was injured and not able to work as a concrete finisher for three days. Because he was able to work in the project office, the superintendent decided not to report the accident.

- Were the actions of the superintendent ethical?
- How would you have handled the situation?

5 Time, Money, and Quality

Learning Objectives

After reading this chapter, you should be able to:

- Describe the ethical challenges that can occur with respect to time, cost, and quality during the management of a construction project.
- Imagine and evaluate potential consequences resulting from ethical decisions made during the management of a construction project.

Introduction

When a contractor signs a construction contract with a project owner, the contractor commits to completing the scope of work defined in the contract documents:

- for the price stipulated in the contract agreement;
- by the date required in the contract; and
- to the quality standards required by the specifications.

The project construction schedule submitted at the start of construction represents the contractor's best estimate regarding the amount of time that will be required to perform each task. However, during the execution of the work, things may occur that will affect the execution of the work necessitating updates to the construction schedule. Periodic schedule updates may be required by the construction contract, or may be required whenever the projected completion date changes.

Just as with the schedule, the cost of completing the project may increase. The cost increases may be due to:

- errors in estimating the cost;
- reduced worker productivity;
- mismanagement of subcontractors; or
- increase in the scope of work.

Scope increases typically are addressed by change orders, while the other cost increases typically are covered by the contractor in a stipulated sum contract or by the project owner in a cost-plus contract. This is because the contractor agrees to complete a specific scope of work for a fixed price in a stipulated sum contract, whereas direct project costs generally are reimbursable in cost-plus contracts.

Material testing requirements and other quality standards are contained in the contract specifications. Some tests may be performed by the contractor, while others may be performed by the project owner or a third-party testing agent. Documentation of all tests is required to be submitted to the project owner. In addition, the owner or her representative usually conducts frequent quality inspections as part of a quality assurance program.

Introductory Case Study

Capital Construction Company has a contract for the construction of a middle school. The contract stipulates that substantial completion must be no later than August 15 to enable the school to move into the building prior to the start of the school year at the beginning of September. If the school is not completed on time, the contract states that liquidated damages will be assessed at the rate of $1,000 per day after August 15. During the procurement of materials for the project, the contractor failed to order needed wall covering for the library and the correct number of interior doors. The contractor's project manager updated the construction schedule for the project to determine the impact of the late delivery of these needed materials. He determined that the impact would be a two-day delay of the completion of the project. He reviewed the schedule and determined that interior painting was on the critical path for the project. To enable the project to be completed on time, the project manager contacted the painting subcontractor and told them that they needed to complete the interior painting in 12 days rather than the 14 days currently scheduled at no additional cost. Was the project manager treating the painting subcontractor ethically?

When preparing the monthly application for payment to the project owner, Capital Construction's project manager reviewed the payment requests that had been received from the subcontractors and the status of work being performed by Capital's workforce. One of the payment items, cast-in-place concrete, being constructed by the contractor's crews seemed to be about 50 percent complete. The site work subcontractor's request for payment indicated that the site work was 70 percent complete. Wanting to minimize negative cash flow on the project, the project manager decided to submit a request for payment to the project owner indicating that the concrete work was 65 percent complete and that the site work was 85 percent complete. Were the actions of the project manager ethical?

While inspecting the installation of the floor covering in the classrooms, the contractor's superintendent noticed that the floor covering installed was not certified as inflammable as required by the contract specifications. Observing that the floor covering was installed properly and looked good, the superintendent decided not to determine whether or not the floor covering material was inflammable. Were the actions of the superintendent ethical?

During a weekly inspection, the contractor's superintendent noticed that mechanical pipes suspended from the ceiling were not installed in conformance with the local building code. The installed pipe looked safe and would later be covered with suspended ceiling. Because the mechanical work was behind schedule, he decided to accept the work and not notify the mechanical contractor to correct the work. Were the superintendent's actions ethical?

Ethical Challenges

Ethical Challenge: Schedule Management

Most construction contracts specify a date by which the construction contractor must achieve substantial completion, which is when the project can be used for its intended purpose even though some work remains to be completed. Failure to achieve substantial completion by the specified date often subjects the contractor to payment of liquidated damages for each day that they are late in achieving substantial completion. To avoid liability for liquidated damages, the contractor develops an initial construction schedule showing completion of the project by the required date. The schedule shows:

- the tasks to be performed;
- the length of time to perform each task;
- the sequence in which each task is to be performed; and
- milestone dates for the completion of the project.

When events occur that impact the completion of the individual construction tasks, these impacts need to be reflected correctly on the schedule so that the impact on the overall schedule for completion of the project can be assessed. Sometimes schedule updates are manipulated to incorrectly indicate the causes of schedule delays. The result may be to incorrectly reflect the causes and impacts of delays in the individual tasks. This manipulation of the schedule is considered unethical.

Ethical Challenge: Progress Payment Requests

Most construction contracts require the submission of applications for payments monthly near the end of the month for the work performed during the month. The general contractor's payment application also includes the payment requests from those subcontractors who worked on the project during the month. The project owner reviews the applications for payment and reviews the status of the project. Based on her review of the project, the project owner's project manager either verifies that the indicated amount of work is correct or engages in a discussion with the general contractor to reconcile the differences. Once agreement has been achieved regarding the amount of work that has been completed, the project owner makes payment to the general contractor, who in turn makes payments to the subcontractors. Since the measurement of the actual work performed during a month may be subjective, there are opportunities for subcontractors or general contractors to bill for work that has not been completed. This is known as overbilling and is considered unethical.

Ethical Challenge: Quality Management

The construction contract specifications contain quality requirements regarding the types of materials and products that are acceptable for the project, the standards for their installation, and the quality of workmanship expected. The submittal process described in Chapter 3 is used to verify that materials and products proposed for the project conform to contract requirements. Quality management involves inspection of project documentation, conducting testing when required, and inspecting the construction work as it progresses. Any discrepancies noted during the inspections generally are placed in a database for tracking until corrected. Follow up inspections are made to ensure that all discrepancies are corrected. Failure to comply with construction quality standards is considered a failure to comply with contract requirements and is considered unethical.

Ethical Challenge: Code Compliance

Many code requirements are incorporated into construction contract specifications by reference. These may include building codes, electrical construction codes, mechanical construction codes, fire protection codes, and occupational safety codes. This is the same as if all of the code requirements were listed in the contract specifications. Compliance with all code requirements is the responsibility of the general contractor. Municipal requirements will specify which inspections must be performed by municipal inspectors during the various phases of the construction project. It is both unethical and in violation of contract requirements to knowingly install improper materials or equipment or to install materials in an improper manner. Upon discovery by inspectors, non-conforming work will be required to be removed and replaced at the contractor's expense.

Safety inspectors may visit the project unannounced to verify compliance with all safety requirements. Failure to comply with required safety requirements usually results in citations and potential fines. Contractors are legally and morally responsible for the working conditions on the project site. Failure to enforce good safety practices is considered unethical.

Applicable Standards

The applicable standards are to practice good faith and fair dealing when addressing time, money, and quality issues during the execution of construction contracts. General contractors need to be fair with subcontractors and project owners when:

- managing their construction schedules;
- requesting change orders;
- submitting applications for progress payments;
- ensuring that quality products and materials are used;
- ensuring that the completed work conforms to quality requirements of the contract; and
- ensuring compliance with applicable codes.

A contractor's professional reputation is greatly influenced by project owners' perceptions of how the contractor handles these important issues. The contractor's performance regarding these aspects of contract management may affect whether or not the contractor will have future opportunities to work with the project owner on other projects.

Construction Participant Perspectives

When project owners select general contractors for their construction projects, they expect the contractors to treat them fairly and to fully comply with all of the terms and conditions of the construction contracts. Likewise, subcontractors expect to be treated fairly when invited to participate on a project. If they perceive that they might not be treated fairly, they may decline the contractor's request to submit a quotation for the work. It is very important for general contractors to maintain good relations with quality subcontractors because their willingness to participate on future projects may significantly affect the success of those projects.

The construction schedule is used to manage the execution of the project. The project owner expects that schedule updates accurately depict the impacts of changes that have occurred during the execution of the project. Subcontractors use the schedule to plan their work, so they also expect it to be current and that the milestone dates depicted will occur as scheduled.

Project owners expect that monthly requests for progress payments are completed accurately and that they are based on the quantity of work completed during the period covered by the request. Subcontractors expect to be paid as soon as the general contractor receives payment from the project owner. Failure to pay subcontractors promptly may cause them financial stress and economic hardship.

Project owners expect general contractors and their subcontractors to perform all quality control measures required by the contract and to ensure that all work performed meets quality and code requirements. This means that contractors and subcontractors must be knowledgeable of the requirements of all codes listed by reference in the contract specifications.

Questions and Scenarios for Discussion

Evaluate the following scenarios.

1 A contractor has a contract for the construction of a large retail store located within a shopping complex. The project owner is very interested in getting the store completed by mid-October to enable the store to open in early November to serve holiday shoppers. The contract contains a provision for $2,000 in liquidated damages for each day beyond October 15 that the contractor achieves substantial completion. During the excavation for the installation of utility lines for the project, the contractor encountered an unanticipated buried water line, which will require additional time to work around. The construction schedule for the building showed five days of float for the utility line construction, and the estimated time for the additional work was three days. Thus the extra work associated with the buried water line would not extend the duration of the project. Because two activities on the critical path were behind schedule, the contractor's project manager changed the logic of the construction schedule to show that the utility line construction was a critical activity and requested a three-day extension to the required project completion date to address the work associated with the buried water line.

   • Were the actions of the project manager ethical?
   • How would you have handled the situation?

2 A contractor has a contract for the construction of a parking garage that contains significant quantities of cast-in-place concrete. The contract specifications require that concrete cylinders be made and tested to verify the strength of the concrete mix. Three samples were taken when placing the concrete for the second floor deck slab and sent to an independent testing laboratory. Two of the samples met the contract specifications regarding strength, and one did not. Consequently the contractor's project engineer decided to provide the project owner only the test results for the two samples that met the contact requirements.

- Were the actions of the project engineer ethical?
- How would you have handled the situation?

3 Cascade Construction has a contract for the construction of a medical office building. The company's project manager decided to hire a subcontractor for the installation of drywall on the project. The drywall subcontractor has been lax at ensuring that all of their workers wear proper personal protective equipment. Cascade's superintendent observed that some of the drywall subcontractor's workers were not wearing gloves and hard hats on the project, but made no attempt to correct the situation even though Cascade's accident prevention plan for the project requires the wearing of both gloves and hard hats.

- Were the actions of the superintendent ethical?
- How would you have handled the situation?

4 Central Constructors has a fixed-price contract for the construction of a community college classroom building. The company's project manager was developing the monthly application for payment to be submitted to the project owner. All of the subcontractors except the mechanical subcontractor had submitted their applications for payment to Central Constructors. Even though the mechanical subcontractor did not submit an application of payment, the project manager planned to include the mechanical work in the application for payment to be submitted to the project owner. The project manager estimated that about 20 percent of the mechanical work was completed during the month. Wanting to help Central Constructor's cash flow on the project, the project manager decided to indicate that the mechanical subcontractor had completed 35 percent of the mechanical work in the application for payment submitted to the project owner.

- Was the project manager being ethical when he billed for the mechanical scope of work when no request for payment was received from the mechanical subcontractor?

- Was the project manager being ethical when he billed for 35 percent of the mechanical work when he believed that only 20 percent of the mechanical work had been completed?

5 Acme Construction has a contract for the construction of a hospital. The company's project manager subcontracted both the electrical and mechanical scopes of work to separate subcontractors. The project superintendent developed a three-week look ahead schedule which showed that the electrical subcontractor could start to work on the project on September 20 and told the subcontractor to start work on that day. The electrical subcontractor arrived with their workers on September 20 and found that the project was not ready for electrical rough-in to start. The project site was not ready for the electrical subcontractor to start work until September 24. Being delayed, the electrical subcontractor submitted a claim to Acme for the additional cost incurred by the delay. Acme's project manager took the claim, added mark-up for Acme, and submitted it to the project owner for payment.

- Were the actions of the project manager ethical?
- What would you have done in this situation?

6 Codes of Conduct, Compliance, and Reporting

Learning Objectives

After reading this chapter, you should be able to:

- Identify the main components of an effective ethics and business conduct program.
- Explain the difference between ethics and compliance.
- Give examples of common core values.
- Give examples of common areas of primary ethical or legal risk.
- Give examples of questions you might ask yourself in an "ethics check."
- Describe a strategy to prevent retaliation against a whistleblower.

Introduction

The construction industry has long struggled with an image of a cutthroat, competitive business environment where award of work to the lowest bidder and low profit margins encourage fraudulent and unethical practices. Forward-looking companies realize that while such practices may pay short-term rewards, they are devastating to long-term success and profitability. As a result, more companies are devoting time and resources to improving their image as responsible and trustworthy business partners. They are accomplishing this by developing codes of ethical conduct and establishing compliance programs to ensure legal operations. In particular, the U.S. government has taken a leadership role in setting standards and guidelines for legal and ethical practices by government contractors. Even in the private sector, studies by consulting firm FMI, in conjunction with the Construction Management Association of America (CMAA), have shown that the number one factor influencing an owner's choice of construction manager or contractor is trust and integrity.

In 1986, some of the nation's top defense companies, including General Electric, Boeing, and Raytheon, came together to establish the Defense Industry Initiative (DII). The Packard Commission, appointed by President Ronald Reagan, had just published recommendations to address widespread allegations of fraud and criminal misconduct in defense contracting. At the time, according to a report by the Ethics Resource Center, the general public had become incensed by reports of "the falsification of timecards and test results, poor quality controls, defective pricing, waste, fraud and overall mismanagement of defense contracts."

Member firms of the DII pledged to support and promote the following six principles of business ethics and conduct.

- Have and adhere to a written code of business ethics and conduct.
- Train employees concerning their personal responsibilities under the code.
- Create a free and open atmosphere that allows and encourages employees to report violations without fear of retribution.
- Monitor compliance, disclose violations, and report on corrective action taken.

- Share best practices and lessons learned with other firms.
- Be accountable to the public.

Beyond legal requirements, DII forums have functioned as environments for sharing best practices that firms can adopt to improve organizational conduct. DII member organizations voluntarily develop written codes of business ethics and share approaches for training employees to comply with the codes.

Starting in December 2007, the federal government began requiring all contractors awarded jobs valued at more than $5 million to have a written code of ethics and compliance within 30 days of being selected for a job. The requirement is spelled out in Title 48 of the Code of Federal Regulations, Contractor Code of Business Ethics and Conduct. It applies to contracts for services and goods involving the General Services Administration (GSA), Department of Defense (DOD), or National Aeronautics and Space Administration (NASA). The regulations do not stipulate what exactly needs to be included in the code of conduct, allowing contractors to develop a code that is appropriate for the size and goals of their company. Many states and private owners also require contractors to have a code of ethics.

Under the federal regulations, a contractor who is required to have a written code of conduct, and a responsibility to discover and report misconduct, may face suspension and debarment by failing to disclose violations of the code. Both suspension and debarment are stiff penalties that refer to ineligibility to compete for federal contracts. Suspension occurs during an investigation or litigation, while debarment is typically for a fixed period of time. This has serious implications, and can substantially impact a contractor's ability to obtain work.

In 2008, a group of large U.S. construction firms recognized the need that construction companies might have to create the newly required code of conduct, and founded the Construction Industry Ethics and Compliance Initiative (CIECI), modeled after the DII. CIECI is a non-profit association of companies within the U.S. construction industry that are committed to the highest level of ethics and conduct and compliance with the law. To assist companies in creating their own codes of conduct, CIECI adopted the DII's toolkit and created a template, available for free download from their website: *Blueprint for Creating and Maintaining an Effective Ethics & Business Conduct Program.*

Even if an owner does not specifically require a contractor to have a formal ethics and compliance program in place, a well-written and fully implemented code of ethics can be a useful tool for marketing and recruiting. It is visible evidence of a company's commitment to its core values and establishing a culture of ethics throughout the company. That level of commitment gives owners confidence, and attracts future employees who wish to work within an ethical culture where they can feel assured their rights and interests are valued.

Introductory Case Study

A contractor has just hired an entry-level estimator to assist with preparing bids. He is assigned to a team under the supervision of a senior estimator. His first assignment is to obtain bids from insulation subcontractors. The senior estimator instructs the new employee to tell each sub who supplies a quote that the contractor already has a number that is 10 percent lower. The junior estimator doesn't think this is right, but he's afraid of losing his new job. How can a company avoid an incident like this? How does a company create a culture of ethics? How might a new employee learn what to do when confronted with a situation that goes against what he or she believes is right? In the following section, we will explore how companies develop codes of conduct that can guide employees in making ethical decisions.

Components of a Business Ethics and Conduct Program

Business codes of conduct generally cover two topical areas: ethics and compliance. A simple way to distinguish between these two is to remember that compliance is based on laws and regulations while ethics is based on values and judgment. The main goal of compliance is to prevent fraudulent and illegal actions, while the main goal of an ethics program is to emphasize the core values that a company wants to be known for and create a corporate culture that embodies those values and helps guide employees when there are no legal rules to go by.

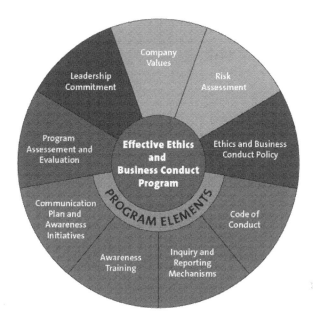

Figure 6.1   Elements of an Effective Ethics and Business Conduct Program. Used with permission from the Defense Industry Initiative.

Figure 6.1 illustrates the components of an effective ethics and business conduct program, as developed by DII and adopted by CIECI.

The foundation of the program is the company's core values, shown at the top of the wheel. Each wedge of the wheel builds on the core values in a clockwise direction, until coming full circle, back to the top. These components are discussed further in the following paragraphs.

Company Values

In Chapter 8, Table 8.1 presents a comparison of the core values of companies selected from the Ethisphere Institute's *World's Most Ethical Companies*. Many of these, as well as additional common core values, are included in the CIECI *Blueprint*:

Honesty	People	Accountability
Integrity	Fairness	Cooperation
Respect	Commitment	Teamwork
Trust	Diversity	Loyalty
Good Citizenship	Leadership	Excellence

Responsibility	Openness	Creativity
Customer Satisfaction	Courage	Dignity
Quality	Safety	Conscientiousness

A company's values statements help define what a company stands for and what they want to be known for. They also make up the foundation upon which a company-wide culture of ethics is built. The core values should be built into every policy and procedure, so that day-to-day internal and external transactions reinforce the company's commitment to ethical practice.

Identifying Risk

Once the foundation of the company core values has been set, we move clockwise on the wheel to Risk Assessment. This is where the company identifies the areas of primary risk to the business. Areas of primary risk often have clearly defined legal boundaries, but can just as often include gray areas where the ethical decision is not obvious. Some examples of risk particular to the construction industry may include:

- Time charging
- Company confidential information
- Competitor information
- Conflicts of interest
- Discrimination and harassment
- Electronic communications
- Environmental protection
- Offering gifts, gratuities, and entertainment
- Accepting business courtesies and kickbacks
- Political contributions and activities
- Quality and safety of the installed work
- Protection and use of company assets
- Security
- Workplace safety.

Each item on the list carries with it a risk of violating one or more of the common core values. For example, if one of your company's core values

Table 6.1 Code of Conduct Contents Selected from CIECI Member Companies

Granite Construction	*PCL*	*DPR*
• Core Values • Our Compliance Program and Guidelines • Standards of Conduct in Business Transactions • Construction Business Standards • Conflicts of Interest • Accident & Injury Prevention • Equal Employment Opportunity (EEO) and Other Employment Laws • Environmental • Copying Documents & Software and the Use of Electronic Media • Trade Secrets & Company Information • Government Contracting • Public Affairs • Antitrust Laws & Competing Fairly • Securities Laws • International Business • Whistleblower • Getting Help • Discipline *www. graniteconstruction.com*	• Our Core Values • Our Commitment • Obey the Law • Act Ethically • Promote a Positive and Ethical Work Environment • Avoid Conflicts of Interest • Recordkeeping • Public Disclosures • Adhere to all Competition and Antitrust Laws • Political Contributions and Activities • Bidding, Negotiating and Performing Contracts • Gifts, Gratuities and Business Courtesies • Doing Business with U.S. Governments • Employment of Government Personnel • Consultants, Agents and Representatives • Protect Proprietary and Confidential Information • Use of Assets • Use of Electronic Communications • Report Unethical Conduct • Cooperate in Ethics Investigations *www.pcl.com*	• Conduct all business with the highest standards of honesty and fairness. • Follow the letter and spirit of the law and uphold all contractual agreements. • Maintain a culture where doing the right thing is not only professed, but prized and practiced by all employees. • Avoid conflicts of interest and circumstances that may lead to even the appearance of a conflict. • Create a safe workplace and uphold a commitment to environmental responsibility. • Exercise common sense and good judgment. *www.dpr.com*

is diversity, incidents of discrimination or harassment in your workplace erode the full realization of that value. Identifying risk means understanding where poor decisions can compromise your company's core values, weaken your reputation, or even lead to illegal actions.

Policies, Procedures, and the Code of Conduct

Each area of potential risk should be addressed by policies and procedures. The company code of conduct is an effective way to organize these
company policies into a useful resource for all stakeholders. Policies,
procedures, and the code of conduct are represented by the next wedges
on the wheel. Codes of conduct do not need to be complicated or lengthy
to be effective. Their purpose is not to dictate every decision, but to
provide guidance to help employees navigate the gray areas presented
by common ethical dilemmas. A code of conduct can also serve as an
effective communication tool to external stakeholders such as clients,
owners, partners, subcontractors, and suppliers, who are also likely to be
affected by the areas of primary risk.

Table 6.1 compares the content outlines of the codes of conduct of
three different member companies of the Construction Industry Ethics
and Compliance Initiative. (Each complete code of conduct is available
for download from the company website.) The comparison illustrates
different approaches companies might take to identify areas of potential
risk and develop policies to address that risk. It also demonstrates that
no single approach fits every company, although there are many common
elements. Company codes of conduct are influenced by company culture,
as well as factors such as whether the company is publicly held or privately
owned, key market sectors, predominance of public or private clients,
relationship with the company's board of directors, and many more.

It can be useful to pick out similar topics among codes of conduct
to learn about areas of primary risk where ethical dilemmas commonly
occur. In Table 6.1, all three of the codes of conduct have a main heading
dedicated to conflicts of interest. Other similarities can be found if you
dig more deeply into the material included under the main headings. For
example, both Granite and DPR have headings that refer to safety and
environmental protection. The PCL code of conduct also covers these
topics, but they are included in the details under the heading "Promote
a Positive and Ethical Work Environment." The organization of the code
of conduct grows out of the manner in which each company wishes to
communicate its ethical priorities.

Handling Inquiries and Reports of Misconduct

Once employees learn to recognize situations that may present risk of illegal action or unethical behavior, they need to know what to do if they find themselves in such a situation, or how to report something they may be witness to. They also need to be reassured that they will not suffer punishment or retaliation for reporting. This is a critical part of an ethics program, and it is why it is specifically addressed in most company codes of conduct.

In Granite Construction's code of conduct, there is a section titled "Whistleblower." A whistleblower is someone within a company who discloses misconduct or illegal action by that organization. If not protected, whistleblowers can be the target of numerous forms of retaliation by company management or by coworkers, from verbal abuse, denial of raises and promotions, or assignment to undesirable locations or tasks, up to physical threats and termination. Even the perception of retaliation can cause a breakdown in trust and disrupt the company culture. Effective codes of conduct address the threat of retaliation by providing a protected means of reporting violations and concerns. The Granite code of conduct explicitly states that whistleblowers are protected from "harassment, retaliation, or any adverse employment consequences" for raising concerns or reporting violations of the code of conduct.

An anonymous or confidential email or telephone hotline provides a safe avenue for employees to voice concerns. Ideally, the hotline is a two-way street, serving as a source of information for conduct-related inquiries, as well as a point of contact for complaints or reports. Calls or emails may be directed to an assigned individual in a senior management position, a company ethics committee or task force, or a person or group on the company's board of directors. Inquiries should be handled by an experienced professional who has the knowledge and the appropriate authority to respond. There should be established procedures in place to ensure that inquiries and reports are handled in a consistent manner. Each inquiry must receive a response, or the program risks losing credibility.

For compliance with federal contract requirements, each call and the action taken must be documented. Regardless of whether the company is engaged in federal or private contracts, all reports should be investigated to determine whether the reported action is a violation of law or company policy. Tracking the number and frequency of complaints or reports can provide valuable information regarding areas of primary risk, and can even reveal problem areas that might have been overlooked. Tracking and reporting are important for assessing the performance of the conduct program, and should receive sufficient and ongoing budget allocation to be effective.

Box 6.1   A Framework for Ethical Decision Making

Recognize an Ethical Issue

Ask yourself:

- Could this decision or situation be damaging to someone or to some group?
- Does this decision require me to choose "the lesser of two evils"?
- Am I rationalizing the situation by telling myself "everyone is doing it"?
- Is this issue more complicated than just doing what's legal?

Get the Facts

Ask yourself:

- What do I know and what don't I know about the situation? What assumptions have I made? Can I find out more? Do I know enough to make a decision?
- What individuals and groups have an important stake in the outcome? Are some concerns more important? Why?
- What options do I have to take action? Have all the relevant persons and groups been consulted? Have I identified all my options? What might I have forgotten or overlooked?

Evaluate Alternative Actions

Ask yourself:

- Which option will produce the most good and do the least harm? (The Utilitarian Approach)
- Which option best respects the rights of all who have a stake? (The Rights Approach)
- Which option treats people equally or proportionately? (The Justice Approach)
- Which option best serves the community as a whole, not just some members? (The Common Good Approach)
- Which option leads me to act as the sort of person I want to be? (The Virtue Approach)

Make a Decision and Test It

Ask yourself:

- Considering all these approaches, which option best addresses the situation?
- If I told someone I respect which option I chose, what would they say?
- If people in my community saw a news story about the option I chose, what would they say?

Act and Reflect on the Outcome

Ask yourself:

- How can my decision be implemented with the greatest care and attention to the concerns of all stakeholders?
- How did my decision turn out and what have I learned from this specific situation? What would I do differently next time?

This framework for thinking ethically is the product of dialogue and debate at the Markkula Center for Applied Ethics at Santa Clara University. Primary contributors include Manuel Velasquez, Dennis Moberg, Michael J. Meyer, Thomas Shanks, Margaret R. McLean, David DeCosse, Claire André, and Kirk O. Hanson. It was last revised in May 2009. Used with permission.

Education, Training, and Communication

An effective code of conduct is a tool to raise awareness of ethical issues and guide employees in making ethical decisions. A culture of ethics is not created by a slogan, or a one-time memo or training session. Employees must first be aware that a code of conduct exists, they must become familiar with the issues and the alternatives, and the information must be readily and easily accessible. Awareness comes from regular reminders in the form of notices, emails, posters, handouts, or weblinks. Familiarity comes from live or online training sessions, reading assignments, and quizzes. The use of real-life scenarios and role-playing activities can help employees practice ethical decision-making skills and become comfortable using the code as a reference.

It's widely known that people have different learning styles. Offering training materials in a variety of formats is a successful strategy to connect with the wide range of educational levels and skills in a typical organization. Because there are so many different ethical issues and situations that can arise within the construction industry, and in conducting business in general, it's not possible or even advisable to attempt to provide employees with an encyclopedia of solutions. Instead, people need to learn the process of recognizing and evaluating ethical dilemmas.

The Markkula Center for Applied Ethics at Santa Clara University has created a useful framework for identifying and analyzing ethical issues. The framework encourages a role-playing approach by leading the inquirer through a series of first person questions. Questions about possible alternative actions are related to the various ethical theories that were discussed in Chapter 1 of this book. Refer back to Chapter 1 if you wish to refresh your knowledge about these ethical theories.

In Table 6.1, we compared the contents of three company codes of conduct. These three companies also offer ethics checks to their employees, consisting of a simple series of questions to ask when faced with an ethical dilemma (Table 6.2).

Table 6.2   Ethics Checks Selected from CIECI Member Companies

Granite Construction	PCL	DPR
As an aid, use the following Ethics Check as a guide during decision making: • Is your behavior/ proposed action legal? Does it comply with the law and Company policies? • Is your behavior/ proposed action something you would like to see published in the newspaper? • Is your behavior/ proposed action something you could comfortably explain to your children? • In short, will your behavior or decision allow you to look in the mirror and feel proud about what you are doing?	When confronted with a situation which raises a concern, ask yourself: • Are my actions legal? • Am I being fair and honest? • Will my actions stand the test of time? • How will I feel about myself afterwards? • Would I think that others were acting unethically if they acted this way? • How would it look in the newspaper? • Will I sleep soundly tonight? • What would I tell my child to do? • How would I feel if my family, friends and neighbors knew what I was doing?	If after reading The DPR Code, you are still in doubt about whether an action was or a potential action is ethical, ask yourself the following questions: • Will the action violate either the law or a company policy? • Will the action damage or be unfair to any of the parties involved? • Will the action make me feel ashamed or uncomfortable looking at myself in the mirror, describing it at a staff meeting, explaining it to my family, or reading about it in a newspaper? If the answer to any of these questions is "yes" or "maybe," you have likely identified an issue that must be either avoided or reported.
www.graniteconstruction.com	*www.pcl.com*	*www.dpr.com*

Measuring Effectiveness

A familiar saying in business is that you can't manage or improve something that you don't measure. In other words, unless you have some way to measure how effective your ethics program is, you won't know if it's working. As we continue to move around the wheel of our Effective Ethics and Business Conduct Program, we reach the wedge for Program Assessment and Evaluation. Earlier, we discussed the importance of recordkeeping and tracking to identify any common themes and ensure that corrective action is being taken. It is also important to evaluate the effectiveness of the methods being used to communicate with and train employees on the contents and use of the code of conduct. Surveys and focus groups are two widely used ways to perform this evaluation. Ethics program effectiveness can be evaluated internally by an assigned group or task force, or an independent outside consultant can be engaged to remove any concerns about potential conflicts of interest.

Commitment from the Top

Ultimately, as we return to the top of the wheel, we come to the need for a visible and ongoing commitment from the company's top leadership to creating an ethical culture within the organization. Ethical leadership can be visible in all the day-to-day operations of a company, from hiring decisions and the manner in which employees are treated; to the type of work pursued; to relationships with clients, subcontractors, and the community; to the prominence assigned to core values and corporate governance on the company website. Leadership commitment to ethical conduct is essential to an organization's reputation for integrity and long-term success for all stakeholders as opposed to short-term rewards and personal gain for a few. Leaders set the example for the rest of the company. Ethical failures within an organization can be traced to a lack of commitment at the top level to making ethical conduct a priority.

The Role of Professional Associations

The construction industry has a wealth of professional associations that provide networking and professional development opportunities, guidance and resources relevant to the many specialties that exist within the sector, including construction managers, cost engineers, estimators, subcontractors, and numerous specialty trades. The majority of these associations have developed their own codes of conduct or ethics to guide their membership. Whereas a company code of ethics is built upon the core values and areas of primary risk specific to that organization, a code of conduct developed by a professional association assists in defining ethical behavior for an industry.

Table 6.3   Codes of Ethics from Selected Professional Construction Associations

American Institute of Constructors (http://www.professionalconstructor.org)

- A Constructor shall have full regard to the public interest in fulfilling his or her responsibilities to the employer or client.
- A Constructor shall not engage in any deceptive practice, or in any practice which creates an unfair advantage for the Constructor or another.
- A Constructor shall not maliciously or recklessly injure or attempt to injure, whether directly or indirectly, the professional reputation of others.
- A Constructor shall ensure that when providing a service which includes advice, such advice shall be fair and unbiased.
- A Constructor shall not divulge to any person, firm, or company, information of a confidential nature acquired during the course of professional activities.
- A Constructor shall carry out responsibilities in accordance with current professional practice, so far as it lies within his or her power.
- A Constructor shall keep informed of new thought and development in the construction process appropriate to the type and level of his or her responsibilities and shall support research and the educational processes associated with the construction profession.

Construction Management Association of America (http://cmaanet.org)

- Client Service. I will serve my clients with honesty, integrity, candor, and objectivity. I will provide my services with competence, using reasonable care, skill and diligence consistent with the interests of my client and the applicable standard of care.
- Representation of Qualifications and Availability. I will only accept assignments for which I am qualified by my education, training, professional experience and technical competence; I will assign staff to projects in accordance with their qualifications and commensurate with the services to be provided; I will only make representations concerning my qualifications and availability which are truthful and accurate.
- Standards of Practice. I will furnish my services in a manner consistent with the established and accepted standards of the profession and with the laws and regulations which govern its practice.
- Fair Competition. I will represent my project experience accurately to my prospective clients and offer services and staff that I am capable of delivering.
- Conflicts of Interest. I will endeavor to avoid conflicts of interest; and will disclose conflicts which in my opinion may impair my objectivity or integrity.
- Fair Compensation. I will negotiate fairly and openly with my clients in establishing a basis for compensation; I will charge fees and expenses that are reasonable and commensurate with the services to be provided and the responsibilities and risks to be assumed.
- Release of Information. I will only make statements that are truthful; I will keep information and records confidential when appropriate and protect the proprietary interests of my clients and professional colleagues.
- Public Welfare. I will not discriminate in the performance of my Services on the basis of race, religion, national origin, age, disability, gender or sexual orientation. I will not knowingly violate any law, statute, or regulation in the performance of my professional services.

▶

- Professional Development. I will continue to develop my professional knowledge and competency as Construction Manager; I will contribute to the advancement of the construction and program management practice as a profession by fostering research and education and through the encouragement of fellow practitioners.
- Integrity of the Profession. I will avoid actions which promote my own self-interest at the expense of the profession; I will uphold the standards of the construction management profession with honor and dignity.

American Subcontractors Association's Model Code of Ethics for a Construction Subcontractor (https://www.asaonline.com)

- Competition. Compete fairly for contracts, avoiding any practice that might be construed to be in violation of the letter or spirit of the antitrust laws. Avoid any activity that could be construed as bid shopping or peddling. Do not knowingly violate any law or regulation governing the competitive process.
- Qualifications. Seek to perform contracts only for projects for which the firm has the technical competence and experience. Do not accept contracts for which the firm is not qualified. Assign staff to projects in accordance with their qualification and commensurate with the demands of the services to be provided under the contract.
- Standards of Practice. Provide materials and services in a manner consistent with the established and accepted standards of the construction industry and with the laws and regulations that govern it. Perform contracts with competence, reasonable care and diligence. Establish prices that are commensurate with the firm's services. Serve customers with honesty and integrity.
- Conflicts of Interest. Endeavor to avoid conflicts of interest, both corporate and individual. Where a corporate conflict exists, disclose such conflict to the firm's customer or prospective customer. Regularly educate staff about personal conflicts of interest and establish a procedure for internal disclosure.
- Public Safety. Assure that the safety of the firm's employees, the employees of others on the job site, and the general public are protected during the provision of the firm's services.
- Service Providers and Suppliers. Treat service providers and suppliers in an equitable manner, assuring that they are provided clear direction and prompt payment for service prodded. Do not knowingly violate any law or regulation governing such relationships.
- Employees. Comply with the letter and spirit of laws relating to working conditions, equal employment opportunities, and pay practices. Do not knowingly violate any law or regulation dealing with employment.
- Public Information. Assure that all public statements and disclosures are truthful. Protect the proprietary interest of the firm's customers.
- Compliance with Laws. Do not knowingly violate any law or regulation.
- Image of the Construction Industry. Avoid actions that promote the firm's own self-interest at the expense of the construction industry and uphold the standards of the construction industry with honor and dignity.

- Internal Procedures. Establish internal procedures under which failure to conform to the above practices will be handled. Each year, review this code of ethics and its internal procedures with each employee. If an employee, customer or other individual becomes aware of a circumstance in which the firm or an employee of the firm fails to conform to the above standards, he/she should immediately report such circumstance to the appropriate individual who will initiate an investigation of and otherwise resolve the reported issue.

Table 6.3 presents the main themes from the codes of ethics of three professional construction associations. Professional association codes can be a valuable resource for companies that are developing their own codes of ethical conduct because they address many of the risk areas inherent in the industry. From the examples provided in the table, we can immediately identify some of the common core values that we have already discussed in this chapter:

- Responsibility
- Honesty
- Respect
- Fairness
- Integrity
- Trust
- Loyalty.

We can also identify some of the common areas of primary risk:

- Company confidential information
- Competitor information
- Conflicts of interest
- Quality and safety of the installed work
- Protection and use of company assets.

Membership in professional associations grants access to a knowledge base of issues, concerns, and policies that are relevant to individuals and companies with common goals and interests. Interaction with other members is useful for obtaining reality checks and for sharing best practices, such as occurs at the annual CIECI forums.

Construction Participant Perspectives

In 2013, Granite Construction was designated one of the World's Most Ethical Companies for the fourth year in a row by the Ethisphere Institute (http://ethisphere.com). The company attributed the repeated accomplishment to its long-standing commitment to the company code of conduct, originally written around 1940 by the company's founder, Walter "Pop" Wilkinson, who named his 11-step document, *Founder's Guide to Future Generations*. The title illustrates the company's desire for long-term success, and the belief that certain core values last through generations. Granite's current Code of Conduct contains an introductory note from President and Chief Executive Officer James H. Roberts, in which he notes the relevance of the original core values, but also emphasizes that the code is a living document, subject to periodic review:

> Our Code of Conduct, while rich with tradition, is regularly reviewed by a team of Granite employees to ensure that it continues to address current challenges and issues. As in the past, our eight Core Values are found to be timeless and appropriate to any situation. Changes in legislation make periodic revisions necessary for the compliance section of the Code.

Later, he makes it clear that, "Under no circumstances should any Granite employee commit an unethical or illegal act under the pretense of being in the Company's best interests." This is an example of a simple statement that provides straightforward guidance and clearly comes from the top level of the company.

Also receiving one of the World's Most Ethical Company designations in 2013 for the fourth consecutive year was the global engineering and construction firm, Parsons. Upon receiving the honor, the company issued this statement describing some of the strategies used to integrate ethics throughout the organization:

Parsons communicates its commitment to integrity through its Code of Conduct, as well as through various training methods, including live training, videos, easily accessed online documentation, and periodic ethics "challenges" or quizzes to keep ethics in the forefront of its daily activities. In addition, Parsons has established metrics to measure the effectiveness of its integrity core value. The corporation also recently established the position of chief compliance officer in order to continue its efforts to incorporate best practices into all aspects of the organization.[1]

As members of the Construction Industry Ethics and Compliance Initiative, all the companies highlighted in this chapter – Granite, PCL, DPR, and Parsons – have demonstrated their commitment not only to instilling a culture of ethics within their organizations, but to sharing best practices with others in the construction industry.

Questions and Scenarios for Discussion

Evaluate the following scenarios.

1 You are working on a proposal for a large project with an important privately owned client. The award will be based on a number of factors, including the qualifications of the proposed management team. A close friend of yours works for a competing firm, which is also submitting a proposal for the same project. One afternoon, while the proposal is still in progress, you meet your friend for lunch. While you are waiting for your food, your friend confides to you that her company has "embellished" their proposed project manager's resume with some false project experience to make him look better. She swears you to secrecy, but she just "had to tell someone."

   • What do you do with this information?
   • Your company's code of ethics includes Service as a core value and emphasizes the interests of your customers. Does this mean you owe it to the client to reveal what your friend told you about your competitor?

- You pass the information on to your manager, and he asks you how you found out. What do you tell him?
- Evaluate the situation using the Framework for Ethical Decision Making. Which ethical theory do you feel gives you the best option?

2 You have been asked to review the entries in your company's Log of Work-Related Injuries and Illnesses. While comparing the supporting documentation to the log entries, you notice that one of the projects had several incidents that were categorized incorrectly. When you check with the project manager, he tells you that there is plenty of leeway in how incidents are categorized and it's standard practice to minimize the severity. You're pretty sure that's not right. But if you correct the log entries, the statistics for the project will look terrible. You're afraid that if you make the corrections, the project manager's performance as well as your company's safety record will be in jeopardy. The project manager has a lot more experience than you do.

- Which is more important for your company's future – some numbers in a spreadsheet, or retaining an experienced project manager?
- Your company's core values include Honesty and Teamwork. If you believe that honesty requires you to correct the log entries, are you still being a team player?
- Evaluate the situation using the Framework for Ethical Decision Making. Which ethical theory do you feel gives you the best option?

3 You work for a company that specializes in storm water control and treatment. As a representative of the company, you have presented several papers about storm water control technology at national conferences, and are considered somewhat of an expert. One of your company's vendors invites you to speak on the first day of a four-day conference in Las Vegas, and is offering to pay for all your travel and hotel expenses for all four days.

- What ethical concerns can you imagine with this scenario?
- Your manager thinks that it would really help your company's

relationship with this vendor if you accepted the offer. Would this be a legitimate reason for accepting?

- Evaluate the situation using the Framework for Ethical Decision Making. Which ethical theory do you feel gives you the best option?

References

Construction Industry Ethics & Compliance Initiative. *Blueprint for Creating and Maintaining an Effective Ethics & Business Conduct Program.* Available for download at http://www.ciecinitiative.org/

Defense Industry Initiative (1986). *Defense Industry Initiatives on Business Ethics and Conduct.* Prepared for the *President's Blue Ribbon Commission on Defense Management*, Appendix M.

Ethics Resource Center, Inc. (1986). *Final Report and Recommendations on Voluntary Corporate Policies, Practices and Procedures Relating to Ethical Business Conduct*, Prepared for the *President's Blue Ribbon Commission on Defense Management*, Appendix N.

FMI Corporation (2011). *Last Line of Defense: The Role of The Hotline in Ethics Programs.* February, 2011.

Note

1 Parsons (2011). Parsons named one of the world's most ethical companies for second year in a row (press release). http:www.parsons.com/media-resources/news

7 Discrimination and Harassment

Chapter Outline

- Learning Objectives
- Introduction
- Introductory Case Study
- Ethical Challenges
- Applicable Regulations and Standards
- Questions and Scenarios for Discussion
- References
- Notes

Learning Objectives

After reading this chapter, you should be able to:

- Describe ways in which discrimination can occur in the workplace.
- Give examples of discrimination on the basis of age, disability, race, and gender.
- Describe the purpose of affirmative action.
- Explain the difference between *quid pro quo* sexual harassment and a hostile work environment.
- Give examples of federal regulations that prohibit various forms of discrimination in the workplace.

Introduction

Jobs in the construction industry are typically described as dirty, dangerous, and requiring physical strength. While there are certainly many tasks that fit this description, there are a great many more that require ingenuity, problem-solving abilities, excellent communication and people skills, and technical knowledge. A narrow view of the work associated with the construction process has in many cases resulted in a narrowly defined, and often discriminatory workforce.

Discrimination in the workplace occurs when someone is restricted or excluded from a job solely because he or she is a member of a certain group, or has physical characteristics or beliefs that differ from the majority. Discrimination often results from stereotyping a group of people and then applying that stereotype to all individual members of that group. The stereotypes usually involve some characteristic, feature, or belief that supposedly makes the group inferior and unfit for employment, promotion, or assignment to certain tasks or a higher level of responsibility. Discrimination can be intentional, or can occur unintentionally based on commonly accepted and unquestioned biases. For example, a company owner may believe that a successful project manager must be aggressive and adversarial, and that only a male employee would have the necessary characteristics to succeed in this position. The company owner may be completely unaware of this bias, and may believe he is assigning the best person for the job. But in this example, he has not only made incorrect generalizations about the skills required for the position, but also about the abilities of his female employees.

Discrimination can be by individuals or by an entire company. Even non-discriminatory policies may have discriminatory outcomes. For example, a company whose policy is to seek hiring recommendations from current employees may inadvertently end up hiring from a narrow pool of people that look and act like the other people in the company. Discrimination can also occur unintentionally in social settings where whoever is present must have the same hobbies or interests as the others in the group to feel like he or she fits in. Someone who doesn't play golf may be left out of discussions and decisions that are made on the golf course. This doesn't mean that these discussions can't occur. It just means that the group must make an effort to seek input from team members who may not be present.

People often fall prey to their misperceptions or biases when making snap decisions or acting on instinct. This is most likely to occur when decisions are made by a single person, in a vacuum, without input or review from others. It's important to consider other options or viewpoints to avoid the pitfalls of knee-jerk reactions, even if this means a decision might not be made as quickly as you would like.

Federal laws prohibit discrimination in employment, but discrimination can also be unethical from several perspectives. From the perspective of the common good, discrimination is unethical because by restricting or excluding certain individuals or groups from equal employment opportunities, it fails to benefit the community of qualified jobseekers. From the perspective of ethics based on rights, the outcome of discriminatory actions in the workplace denies an equal right to employment to members of a particular group. Discrimination also fails the basic application of the golden rule since no one would choose to be discriminated against by others.

Because stereotypes are so common and widespread, the opportunities for discrimination are as well. Discrimination can occur on the basis of assumptions that people make about:

- age;
- gender;
- religious beliefs;
- racial and ethnic groups;
- disability;
- sexual orientation.

Introductory Case Study

Arun wears a turban as required by his Sikh religion. His supervisor tells him that his turban makes his coworkers "uncomfortable" and asks him to remove it while he's at work to avoid harassment and unwanted attention. The supervisor cites a company policy that prohibits employees from wearing hats or other head coverings in the office. Is this a reasonable request, or is this discrimination? Would the request be reasonable if Arun worked on a jobsite where hardhats were required?

Ethical Challenges

From the perspective of virtue ethics, discrimination on any basis violates many of the commonly accepted core values, such as fairness, diversity, and respect. Discrimination also opposes the rights of individuals and the common good of the community. The following paragraphs examine some of the common aspects of various forms of discrimination.

Age Discrimination

As with other forms of discrimination, age discrimination occurs when assumptions are made about the characteristics or abilities of an entire group of people based on stereotypes and without supporting evidence. For older workers, those stereotypes might include reduced mental acuity, unfamiliarity with technology, and physical or health limitations. While these issues might occur with older workers, there is no supporting evidence that justifies the stereotype for all the employees or job applicants in this group. In fact, these same issues can also frequently occur in younger workers. In general, age discrimination falsely assumes that worker productivity decreases with age.

According to research by the Center for Construction Research and Training (CPWR), between 1985 and 2010, the average age of U.S. construction workers (including trade workers, contractors, and construction managers) jumped from 36.0 to 41.5 years old.[1] Many workers choose to delay retirement for financial reasons as the eligible age for collecting Social Security benefits has increased over time, and as economic factors have affected retirement savings. As more older workers enter or choose to stay in the workforce, employers need to be aware of outright or hidden bias concerning the abilities of this group. Although physical strength is frequently required for certain manual construction labor tasks, in many cases, simple adjustments to work activities such as attention to safe practices and ergonomics can reduce the potential for injury and benefit both older and younger workers.

When it comes to discrimination against older job applicants, even normal hiring practices may contain pitfalls that can bring out hidden biases. For example, the simple requirement for job applicants to state years of experience can inadvertently reveal the age of the applicant. It is important during the hiring process to be aware of clues like this that may subtly influence hiring decisions.

The potential for age discrimination should be of concern to everyone in the workforce. Although some employees may escape labeling or stereotyping for most of their careers, everyone, as long as they are living and working, eventually becomes a member of the over 40 age group. As an ethics check, one need only ask oneself, "When I am this applicant's or employee's age, how do I want to be treated?"

Discrimination Based on Disability

Disability discrimination occurs when an employee or job applicant is treated less favorably because he has an existing disability, a history of a disability (such as cancer that is controlled or in remission), or because she is believed to have a lasting physical or mental impairment. Disability discrimination can also occur if an applicant is excluded from consideration because of a relationship to someone with an impairment. For example, an employer may assume that a jobseeker with a disabled spouse or child will be unreliable or require extra time off to care for that family member. If that assumption removes an otherwise qualified candidate from the employment pool, that is discrimination.

Discrimination Against National Origin, Race, Color or Religion

National origin can be defined by, or inferred from an individual's birthplace, ancestry, culture, last name, accent, or native language. It is unlawful to discriminate against any employee or applicant on the basis of that individual's national origin, race, color, or religion. It is also unlawful to deny equal employment opportunity because of marriage or association with other members of a national origin group or because of attendance or participation in schools, churches, temples, or mosques generally associated with a national origin group.

A gray area frequently encountered in the construction industry is whether a company can require that employees be fluent in English. Would this policy discriminate against employees or job applicants who spoke English as a second language or with an accent? According to the U.S. Equal Employment Opportunity Commission (EEOC), a fluency requirement is only permissible if required for the effective performance of the position for which it is imposed. An English-only rule may be adopted only for a non-discriminatory reason, such as being necessary for safe operations. An example of this might be a position responsible for safety training where fluent English and a minimal accent are necessary to ensure the effective communication of critical information. An employer may not base a hiring or other employment decision on an applicant's foreign accent unless the accent poses an obvious interference with job performance.

Discrimination Based on Gender or Sexual Orientation

Although the law against discrimination on the basis of gender or sexual orientation legally pertains to both men and women, the most common occurrence in the construction industry involves discrimination against women. As mentioned in the introduction to this chapter, stereotypes about the behavior and abilities of women in construction careers often lead to their placement in lower-paying jobs and fewer opportunities for advancement.

Women themselves face tough decisions when it comes to balancing job and family responsibilities. Since responsibility for childcare is still most frequently assumed by the female spouse in a traditional heterosexual relationship, it can be difficult for women to advance in their careers if they have taken leave from their jobs to care for young children. Forward-thinking companies are creating policies for flexible work schedules, telecommuting, and alternate work assignments to help retain valued female employees. Although not every company is able, or can afford to implement such policies, employers must be aware that hiring decisions that exclude women on the basis that they will take time off or leave to raise a family are discriminatory.

Hiring and job assignment decisions that discriminate against qualified female candidates may also be made out of ignorance about, and fear of sexual harassment definitions, or reluctance to invest in appropriate workforce training concerning discrimination and harassment policies. Differences, misunderstandings, and confrontations between men and women in the workplace are common occurrences; however, the solution is not a segregated workforce, but rather clearly communicated company policies, a designated individual or office where employees can go to ask questions and voice concerns, and open channels of communication. It is important for contractors to recognize that the diversity of their client pool is increasing, and that a diverse workforce, especially in management positions, will expand their ability to be responsive to a wider range of communication styles and preferences among those potential clients.

Affirmative Action

Although a company may currently have rigorous policies in place barring discriminatory employment practices, there may still be barriers to advancement for underrepresented groups or individuals due to biased decisions that were made in the past. The purpose of an affirmative action program is to create fairness and equal opportunity in organizations where past practices may have resulted in widespread and institutionalized discrimination. This is the most important thing to remember about the purpose of affirmative action. Without it, it is almost impossible for under-represented groups to advance in occupations like construction where leadership positions continue to be dominated by white males. Benefits of affirmative action include increased diversity within the organization, and a larger network of job applicants, subcontractors, and vendors to make the company stronger and more resilient.

There are different levels of affirmative action. The strongest form involves reserving a portion of business exclusively for minorities. This is the case with set asides for disadvantaged, minority, or woman owned businesses. At a less strict level, the company can show a preference for minorities in a mixed pool, or encourage minorities to apply and offer incentives. At its weakest level, an employer may simply choose to advertise in minority publications or associations.

There has been some debate about whether affirmative action programs are themselves a type of discrimination. Some have claimed that affirmative action, in seeking to increase workplace diversity, is a form of "reverse discrimination" against white males. However, as we discussed in the introduction to this chapter, discrimination involves labeling an individual as inferior based on a stereotype of the group to which the individual belongs or is associated with. It cannot be said that an affirmative action program applies any such stereotype of inferiority to white males. A valid concern is the potential abuse of an affirmative action program by individuals falsely identifying themselves as members of a disadvantaged group in order to receive preference in an employment action.

From an ethical perspective, affirmative action employs the utilitarian approach, in which the desired outcome is the greatest good for the greatest number of people. It is also aligned with the common good approach. Although the rights of an individual member of the dominant group (typically a white male) may be violated by an affirmative action decision, the outcome of greater diversity in the workplace benefits the greater good of the organization and the community.

Sexual Harassment

Sexual harassment with respect to the law is defined by the EEOC as:

> unwelcome sexual advances, requests for sexual favors, and other verbal or physical conduct of a sexual nature when submission to or rejection of this conduct explicitly or implicitly affects an individual's employment, unreasonably interferes with an individual's work performance, or creates an intimidating, hostile or offensive work environment.[2]

There are many common misconceptions about the nature of sexual harassment. In reality, the victim as well as the harasser may be a woman or a man. The victim does not have to be of the opposite sex. In fact, data from the EEOC presented in an article in Engineering News-Record (ENR) showed that between 2002 and 2011, 20 to 25 percent of complaints brought against construction companies involved a harassment claim by a man.[3]

The victim also does not have to be the person harassed but could be anyone affected by the offensive conduct. The harasser can be the victim's supervisor, a supervisor in another area, a coworker, or a nonemployee such as a client, subcontractor, or a supplier doing business with the company.

There are two general categories of sexual harassment: *quid pro quo* (literally "this for that") and hostile work environment. With *quid pro quo* harassment, a sexual favor is expected in return for receiving a job benefit such as a promotion or a raise. *Quid quo pro* can also describe a situation in which benefits are denied if an employee refuses a sexual advance. A hostile work environment can be created as a result of unwelcome teasing, physical contact, or offensive comments or pictures. Although the law doesn't prohibit any of these things if they occur as isolated incidents, harassment is illegal when, in addition to being unwelcome, it is also frequent, persistent, or severe.

A concern for victims of sexual harassment is the threat of retaliation for reporting workplace misconduct, either from the harasser, or from other coworkers. For this reason, an important component of sexual harassment training includes educating employees about how to recognize and stop retaliation, and to make it clear that retaliation or further harassment of individuals who raise concerns or report violations will not be tolerated. Chapter 6 (Codes of Conduct, Compliance, and Reporting) presents more information about protection against retaliation.

From an ethical perspective, companies have an obligation to provide a workplace free from sexual harassment, and employees have a right to work in a safe and non-threatening or offensive environment. An excellent example of a clearly stated policy on harassment and retaliation is found in the Code of Conduct for Granite Construction, a member company of the Construction Industry Ethics and Compliance Initiative (CIECI). Granite's policy on Equal Employment Opportunity (EEO) and Other Employment Laws includes these two statements:

> The Company has zero tolerance for discrimination or harassment of any kind, and employees will be subject to disciplinary action, including termination, for violations.

Box 7.1   Hostile Work Environment or Harmless Trash talk? EEOC v. Boh Brothers Construction Company

In this 2011 lawsuit, a former employee of Boh Brothers Construction Company charged that a superintendent harassed and taunted him by engaging in verbal abuse and taunting gestures of a sexual nature, and by exposing himself. The harassment took place in 2006 on the I-10 Twin Span project over Lake Pontchartrain between Slidell and New Orleans, LA. After making a complaint about the harassment, the employee was transferred to another jobsite, and then laid off, allegedly because of a lack of available work. He then filed a charge with the EEOC, complaining of a sexually hostile work environment and retaliation. The EEOC presented evidence at trial that the victim's supervisor harassed him because he thought he was feminine and did not conform to the supervisor's gender stereotypes of a typical "rough ironworker." Following a two-and-a-half-day trial, the federal jury awarded the victim damages for the sexual harassment claim, including punitive damages and emotional distress. The EEOC established that Boh Brothers had no policy that defined or specifically prohibited sexual harassment. The superintendent testified that before this lawsuit, he had never received training on sexual harassment.

A year later, the 5th U.S. Circuit Court of Appeals reversed the decision. The appeal ruling stated that the superintendent's behavior was not indicative of sexual harassment, "nor is it the business of the federal courts generally to clean up the language and conduct of construction sites." A Boh Bros. spokesperson said the company disciplined the superintendent for his behavior, and was implementing a training program on appropriate behavior for all workers. In their ruling, the three-judge panel noted that the superintendent was "a world-class trash talker and the master of vulgarity in an environment where these characteristics abound."

In a follow-up editorial, ENR staff expressed their opposition to the reversal of the decision. The superintendent's unrelenting harassment created a hostile work environment ... that no employee should have to endure. The judges' characterization of the behavior as fun-lovin' horseplay and needling ... misses the whole point of statutory protection for employees. The ruling creates a second-tier workplace category for construction by condoning [the superintendent's] actions as 'non-sexual' trash-talking ...

The editorial also pointed out that prior to the harassment, the victim had declined to join a union amid unionized workers, and was then subjected to name-calling with terms such as "rat." The abuse subsequently escalated to sexual humiliation and attacks on the victim's masculinity.

Sources:
Editorial: Justice Denied in Boh Bros. Same-Sex Harass Verdict. *ENR,*
August 7, 2012.
Judges Overturn Same-Sex Harass Verdict Against Boh Bros. *ENR, July 31, 2012.*
EEOC Obtains $451,000 Jury Verdict Against Boh Brothers Construction Co. for
Male-On-Male Sexual Harassment. *EEOC press release, March 29, 2011.*

The Company will not tolerate retaliation against anyone who in good faith raises a concern or reports a violation.[4]

Applicable Regulations and Standards

The U.S. Equal Employment Opportunity Commission (EEOC) is the federal agency responsible for enforcing federal laws against various forms of discrimination in the workplace. Most employers with at least 15 employees are covered by EEOC laws, as are most labor unions and employment agencies. The laws apply to all types of work situations, including hiring, firing, promotions, harassment, training, wages, and benefits. Some specific examples are described in this section.

In 1964, the federal government took the first big step to address discrimination in the workplace by passing the Civil Rights Act. Title VII of the act prohibits any form of discrimination by any employer with 15 or more employees, on the basis of race, color, religion, sex, or national origin. It is important to note that lesbian, gay, bisexual and transgender individuals may bring sex discrimination claims under Title VII, such as allegations of sexual harassment, or adverse actions resulting from the person's sexual orientation or gender identity.

In 1967, the Age Discrimination in Employment Act (ADEA) was added to protect employees and job applicants aged 40 or older from employment discrimination based on age. Under the ADEA, it is unlawful to discriminate against an older worker when it comes to any employment decision, including hiring, firing, promotion, compensation, benefits, job assignments, or training. Since its inception, amendments to the ADEA have included prohibiting mandatory retirement (1978), eliminating an upper age limit for protection under the act (1986), and enacting the Older Workers Benefit Protection Act (1990). The ADEA applies to companies with at least 20 employees.

The Pregnancy Discrimination Act (PDA) is another amendment to Title VII of the Civil Rights Act of 1964. This amendment makes it unlawful to discriminate on the basis of pregnancy, childbirth, or related medical conditions. An employer cannot refuse to hire a woman if she is pregnant as long as she is able to perform the major functions of her job. Women affected by pregnancy or related conditions must be treated in the same manner as other applicants or employees who are similar in their ability or inability to work. If a woman is unable to continue in a certain position due to pregnancy, the company may wish to offer an equally rewarding and compensated position in a different capacity to allow them to retain a valued employee.

The Family and Medical Leave Act (FMLA) of 1993, enforced by the U.S. Department of Labor, protects employees who must take leave from their jobs for specified family and medical reasons. For example, a new parent (including foster and adoptive parents) may be eligible for 12 weeks of leave for care of the new child. In general, the FMLA applies to private sector companies with 50 or more employees, and any public sector agency, regardless of the number of employees, although there are certain exceptions and restrictions. In most cases, employees must have worked for the company for at least 12 months to be eligible.

The Americans with Disabilities Act (ADA) protects qualified employees or job applicants from being treated unfavorably because they have a disability. The law prohibits discrimination in all decisions related to employment. The employment provisions of the ADA apply to private employers, state and local governments, employment agencies, and labor unions, all with 15 or more employees. (Federal employees and applicants are covered by the Rehabilitation Act of 1973, instead of the ADA.) ADA requires an employer to provide reasonable accommodation to an employee or job applicant with a disability, unless doing so would cause significant difficulty or expense for the employer (referred to as "undue hardship"). The law also protects someone from discrimination based on a relationship with a person with a disability. For example, it is illegal to discriminate against an employee because her husband has a disability.

An employer can claim undue hardship when asked to accommodate an applicant's or employee's disability, religious practices, or other protected situation, if the accommodation would result in any of the following:

- more than ordinary administrative costs;
- another job becoming less efficient;
- an infringement of other employees' job rights or benefits;
- a less safe workplace;
- other workers having to carry an additional share of potentially hazardous or burdensome work;
- a conflict with another law or regulation.

Recognizing that some employers might struggle with the financial burden required to accommodate a disabled employee, the Internal Revenue Service offers certain tax credits or deductions for costs incurred in making a business more accessible, such as providing disabled access, or removing architectural and transportation barriers.

Questions and Scenarios for Discussion

Evaluate the following scenarios.

1   A female project engineer is attending a meeting in a subcontractor's jobsite trailer. This is her first field assignment and she wants to make a good impression and show that she knows her stuff. When she walks into the conference room, she sees several "men's" magazines on the table and a calendar on the wall with a photo of a nude woman. She asks the subcontractor's representative to please remove the magazines and the calendar before the meeting. The sub laughs and tells her that it was her choice to work in a man's job so she better just get used to it.

   - What should the project engineer do next?
   - What is her employer's responsibility?
   - What is the subcontractor's responsibility?

2   Roger has been an estimator for 20 years. He's good at his job, reliable, and has been a good mentor for the younger employees. Recently, Roger applied for a new position as a lead estimator responsible for organizing the effort for new bids and assigning tasks to several junior estimators. One afternoon, Roger accidentally overhears two managers talking about how they need to promote more "younger" people because they have more energy and drive, fewer health issues, and aren't thinking about retirement. Roger is worried that he will get passed over for promotion because of his age.

   - What should Roger do?
   - What is his employer's responsibility?

3   Sunita, who has a noticeable Indian accent, has five years of experience as a front desk receptionist for a local planning department. Through her job, she has met many local contractors and building professionals. A friend of hers, who works for one of the local contractors, told Sunita that her company was looking for a new receptionist, and encouraged Sunita to apply for the job. After her interview, Sunita was told that although she was otherwise qualified for the job, the contractor needed someone at the front desk who "sounded more American" and was "easier to understand."

- Was this a legitimate reason for not considering Sunita for the position?
- Would the contractor have had the same difficulty hiring someone with a British or a Southern accent?

References

For information about different types of discrimination, for employees and employers, and current federal regulations, please visit the website of the U.S. Equal Employment Opportunity Commission at http://www.eeoc.gov/.

For more discussion about ethical issues related to discrimination, an excellent resource is *Business Ethics* by William H. Shaw (2011), published by Wadsworth Cengage Learning.

Notes

1 *The Construction Chart Book: The U.S. Construction Industry and Its Workers, 5th Edition*, CPWR – The Center for Construction Research and Training, April 2013.
2 EEOC, *Facts About Sexual Harassment*, http://www.eeoc.gov
3 *Construction Men Find Refuge in Sexual Harassment Laws*, ENR, August 28, 2012.
4 Granite Code of Conduct, http://www.graniteconstruction.com/

8 Client Relations

Learning Objectives

After reading this chapter, you should be able to:

- Describe ethical challenges that can occur between a contractor and client.
- Explain the difference between a gift and a bribe.
- Imagine and evaluate potential consequences resulting from common ethical decisions in contractor/client relationships.
- Recognize the statutes that regulate relationships with representatives of federal agencies.

Introduction

Project owners are a diverse group. An owner may be a federal, state, or local agency; a publicly held or private corporation; a municipal authority, such as a school district or local transportation authority; a utility company; a non-profit organization; or an individual developer. Owners may have substantial experience in managing construction projects and contracts, or may have none at all. The ideal owner/contractor relationship is built on a foundation of mutual trust; unethical behavior by either side can erode that trust and lead to suspicion, blaming, retribution, justification of further unethical decisions, and possibly legal action.

Consulting firm FMI in conjunction with the Construction Management Association of America (CMAA) conducts annual surveys of construction project owners about their perceptions of and experiences with a variety of topics and issues, including risk management, project delivery, technology, and ethics.

In the 2005 Sixth Annual Survey of Owners, FMI/CMAA asked owners: what characteristics do you use to choose the right construction manager and contractor for your project? Setting aside the cost constraint, the number one factor that influenced the owners' choice of construction manager or contractor was trust and integrity. Other important factors for owners were the contractor's ability to control schedule and cost, experience level, and communication and leadership skills.

Integrity is commonly listed among a company's core values. Table 8.1 shows the core values for four companies who have repeatedly been among those honored by the Ethisphere Institute as the World's Most Ethical Companies. Note that instead of integrity, CH2M Hill lists respect, which is a consequence of acting with integrity. A company's core values provide the framework and the context for making decisions. By listing integrity as a core value, a company is telling its clients, investors, subcontractors, suppliers, and employees that when ethical challenges occur, maintaining the company's integrity will be a guiding principle.

Integrity is often described as "walking your talk." In his book *The Integrity Chain*, author Ralph James lists things that he would do to demonstrate integrity. This list can also be viewed as simple rules for building trust between you and your client.

Table 8.1   Core Values of Companies Selected from the Ethisphere Institute's
World's Most Ethical Companies

Granite Construction	Parsons	CH2M Hill	Fluor
Safety	Safety	Respect	Safety
Honesty	Quality	Delivery Excellence	Integrity
Integrity	Integrity	Employee Control	Teamwork
Fairness	Diversity		Excellence
Accountability	Innovation		
Consideration of Others	Sustainability		
Pursuit of Excellence			
Reliability			
Citizenship			
www. graniteconstruction.com	*www. parsons.com*	*www.ch2m.com*	*www.fluor. com*
Ethisphere Institute: http://ethisphere.com/			

- Show up on time to meetings.
- Follow up on action items.
- Don't make promises you can't keep.
- Act consistently.
- Don't cover up bad news.
- Don't look for scapegoats to take the blame.
- Know your policies and procedures.
- Go through proper channels instead of looking for ways to game the system.

When trust and integrity are weak, owner/contractor relationships can become adversarial. Each may attempt to gain the upper hand by engaging in unethical or even illegal practices. If unethical actions occur at top management levels, there is a risk that these practices may become a way of doing business for that organization. A contractor who doesn't trust an owner to award contracts fairly may try to obtain a preferred status by giving his client a gift or favor (unethical and possibly illegal). An owner who doesn't trust her contractor may withhold payment or cut costs on change orders. A contractor who doesn't receive fair or timely payment may try to compensate by overbilling. Some examples of unethical practices between contractors and clients are presented in this chapter.

Introductory Case Study

A contractor learns that the head of the local Department of Public Works, who signs off on all major procurement for the County, owns a vacation home nearby. This official has many friends in the contracting community who frequently drop by and help out on home improvement and remodeling projects. Some of these contractors have even donated supplies along with their own labor. Our contractor has been invited to a "work party and barbecue" on the upcoming weekend to help build a new fire pit at the official's vacation home. She is concerned that if she doesn't participate, her company will be at a disadvantage when it comes time to bid on Public Works projects. Can our contractor help out at the work party? Can she bring along some of her employees to help too? What other choices might she have?

Ethical Challenges

In the following paragraphs, we will discuss some of the common ethical challenges encountered in contractor/client relationships, beginning with the issue of gift giving, which was illustrated in the introductory case study.

Ethical Challenge: Gift Giving

Particularly when a large client offers the potential for ongoing projects, a contractor may feel a need to set himself apart from the competition somehow, attract attention, or gain favor with that client. Many contractors have sought ways to gain "a foot in the door" with an important client in order to be placed on short lists to be able to bid or propose on work. A contractor can differentiate herself from the competition naturally over time through outstanding performance, earning a reputation for integrity, quality, safety, and value. However in the short term, and in the absence of a relationship that has been built over time, there can be a temptation to look for a short cut.

A personal relationship with an owner or an owner's employee might present an opportunity for either party to take a short cut. There's an important distinction between a friendly working relationship and taking unethical advantage of that relationship to gain a competitive edge or financial reward. A personal relationship between a representative of the owner and a representative of the contractor might even predate their business relationship. Typically, there is no ethical issue when an employee accepts a gift when it is clear that the gift is motivated by a family relationship or personal friendship rather than the position of the employee. However, in this case, both parties must be especially sensitive to perceived impropriety.

Some companies allow gift giving to clients or potential clients. The *ethical* challenge is being able to determine when a gift is acceptable and appropriate. The *legal* challenge is being able to distinguish the difference between a gift and a bribe. The following two conditions transform gift giving to bribery:

1 if the gift predisposes the recipient to favoring the interests of the gift giver;
2 if the value of the gift places the gift giver in a privileged position.

A good definition of a gift is something of value given without the expectation of return. In general, token gifts, meals, and entertainment are allowable if acceptance is not illegal and if the gift is clearly appropriate under the circumstances. In addition, it should also be possible, acceptable, and appropriate for the gift to be reciprocated. That is, a gift should not be given if it is clearly beyond the ability of the recipient to return an equivalent gift. For example, it is typically not unethical to pay for a client's meal if the client has the opportunity and the means to reciprocate at another time.

There are additional important considerations if your client is an employee of a federal, state, or local government, or a public agency. Federal government employees are restricted by law from accepting gifts over certain dollar thresholds. Federal regulations state that no single gift or item to a representative of the federal government should exceed a value of $20, and the annual limit for more than one gift is limited to $50.00.[1] State or local governments may have similar restrictions. Even a token gift as simple as a contractor-branded cap or coffee mug may pose a problem for a public employee. Not only may the gifts exceed the allowable value limits, but a government employee wearing your company's t-shirt or carrying around a mug with your logo could create an appearance of impropriety or conflict of interest.

If you can answer "yes" to any of the following questions, then chances are, the gift is unacceptable.

- Is the gift illegal?
- Is the gift cash?
- Does the gift exceed a modest value?
- Could the gift give the appearance of an unfair business advantage?
- Is the gift prohibited by the recipient's employer?
- Would public knowledge of the gift cause embarrassment or reflect negatively on the individual gift giver, the individual's company, or the recipient?

Box 8.1   Gift or Bribe? United States v. Hoffman

In United States v. Hoffman (2009), the appellate court upheld a gratuities conviction based on an indictment alleging that the defendant had given a government employee a set of golf clubs for or because of that government employee's role in rating the contractor's performance under a contract with the United States Army Corps of Engineers.

Russell Hoffman was a vice president at Surdex Corporation, and project manager for a two-year contract for aerial photography and mapping services that was awarded in 1996. William Schwening was the Army Corps of Engineers' primary point of contact for the project. Schwening was also responsible for completing a performance evaluation report grading Surdex's performance under the contract.

The primary evidence consisted of email correspondence between Hoffman and Schwening in which statements by Schwening somewhat jokingly implied that he was delaying Surdex's performance evaluation until he received a new set of golf clubs. Hoffman subsequently sent Schwening a set of golf clubs that were paid for by Surdex.

> The fact that it was the company that purchased the clubs rather than Hoffman suggested that the gift was given for a business purpose, and not based on a friendship between the two men.
>
> The jury acquitted Schwening on all counts but convicted Hoffman. Although Hoffman appealed, the conviction was upheld because the gratuities statute states that the government need only establish that a gift was given with the *intent* to induce or reward an official act, regardless of whether that act occurred or whether the gift was accepted.
>
> Source: *http://www.governmentcontractslawblog.com*

Ethical Challenge: Bad-mouthing a Competitor

Another short cut around establishing a trustworthy reputation through performance is to offer the client negative or harmful information about the competition. One way this can happen is when employees move from one contractor to another, bringing with them "insider" or proprietary information. Proprietary information that can damage a competitor might include knowledge about upcoming changes in key personnel, financial or legal problems, or issues with jobsite training and procedures. Is it unethical for a contractor to share this kind of information with a client, or is she just looking out for her client's best interest? There are several potential consequences to consider.

Whether you are the contractor thinking about divulging information, or the client who is receiving the tip, it's important to consider the source of the information, how the source came by the information, and the motivation of that source. Both should be asking themselves: how current and accurate is the information? Has it been verified or is it a rumor? The contractor should be asking himself: what do I think I can gain by telling this to the client? The client should be asking herself: why is this contractor choosing to give me this information at this particular time? From the viewpoint of the contractor, the tactic of bad-mouthing the competition is truly a shortsighted approach. Contractual relationships differ from project to project. Today's competitor might be tomorrow's construction manager, or subcontractor, or joint venture partner. Is the contractor willing to end all possible future work with this competitor for a better shot at the current project?

From the viewpoint of the client, knowing that a contractor has revealed confidential information about a competitor may seriously impact the sense of trust that an owner has with that contractor. Revealing information about a competitor may also have the result of the owner requesting additional requirements of all bidders, and thus may backfire on the company revealing the information. In short, it is best practice for all parties to be responsible for their own due diligence concerning their potential clients, contractors, and partners.

Ethical Challenge: When a Client Takes Advantage of the Owner/ Contractor Relationship

In some cases, the owner is well aware that the contractor is motivated to please the client to the extent possible in order to retain favorable status. What happens when an owner takes advantage of this motivation to secure uncompensated work by the contractor? How much free consulting is ethically acceptable and where does the contractor draw the line? What happens if the owner receives information given in good faith by a contractor and then uses that information to enter into a contract with a competitor at a lower price? How does a contractor protect himself from an unethical client? How does this situation change if our contractor is now the recipient of privileged information? How does the contractor act on such information?

Naturally, owners seek to obtain the best ideas and approaches that will make their projects successful. However, once in possession of proprietary information regarding a contractor's means and methods, clients must respect this information as intellectual property and not divulge confidential information from one contractor or architect to another. At the same time, if such information is offered to a contractor or architect for review, the ethical response would be for them to refuse to see it. Some unethical behavior on the owner's side is attributed to inexperience. When inexperienced owners choose to act as their own construction manager, serious issues and misunderstanding can arise.

Ethical Challenge: Overbilling

When the opportunity arises, a contractor may seek to make more money on a project by overcharging on changed work, or overbilling for work that is running under budget. What is the downside for the contractor to overcharging? He may recoup some money, but at what cost to his reputation? Is some overcharging permissible if it is reasonable and fair? If overcharging is allowed or encouraged, what does that communicate to employees about the values of the company?

In another situation, an owner may have an unwritten policy of automatically cutting a percentage of every proposed change order, causing the contractor to respond by padding each change order for negotiation purposes. This approach sends a signal to the contractor that the owner is expecting the contractor to overcharge, and further, that the owner is accepting this as a standard way of doing business.

From the contractor's standpoint, it is unethical for an owner to withhold payment for work done and the contractor is justified in overcharging to maintain a positive cash flow. From the owner's standpoint, the contractor is acting unethically by overcharging for work not done and the owner must withhold payment to compensate. How can this standoff be resolved? Is an ethical code relative, that is, can an action be ethical in one situation and unethical in another?

Table 8.2 shows one way of analyzing this situation. Each possible action is based on a core value and has potential consequences that can be designated as a benefit or a risk. The basis for the decision is whether the benefits of the action outweigh the risks, and if the outcome is aligned with the contractor's core values and code of conduct.

Table 8.2 shows us that if a company's core values, and hence its ethical code of conduct, are based on improving the company's bottom line, then the benefit derived from overcharging does in fact outweigh the risks. However, if the contractor's core values are honesty, integrity, and trust, then overcharging is never aligned with an ethical code of conduct, even if it can be justified from a financial standpoint. If we refer back to the core values listed in Table 8.1, we will not find one value, among any of those listed, that represents monetary gain. It is important to remember that just because an action can be justified, it isn't automatically or necessarily ethical.

Table 8.2   Scenario: Contractor has an opportunity to overcharge for work done on a project

Action	Core value basis	Potential benefits	Potential risks
Do not overcharge	• Honesty • Integrity • Trust	• Enhance reputation in the contracting community • Build goodwill with the owner	• Don't make as much money
Overcharge	• Improve the company's bottom line	• Make more money on the current project	• Damage reputation • Lose current and possibly future clients • Set a precedent within the company for overcharging

For each possible action, do the benefits outweigh the risks?

Some contractors have defined overbilling as rebalancing or front-end loading the payment schedule to place a higher value on work performed early in the project and lower values on work to be completed near the end of the project so that the contractor does not incur a seriously negative cash flow. This is a completely different situation from overcharging for work not done, and should not by all rights be called overbilling. Front-end loading the schedule of values, when limited to a few planned early completion activities, is a common and generally accepted practice that should be negotiated and agreed on by the contractor and the owner. Inexperienced clients may require some education on front-end loading to fully understand why the practice is necessary.

Ethical Challenge: Timely Exchange of Information

Owners recognize that they share in the responsibility for a successful owner/contractor relationship. In the 2005 Survey of Owners mentioned at the beginning of this chapter, FMI/CMAA asked owners what responsibilities they felt they themselves needed to improve. Their top answers were:

- Making timely decisions
- Requiring good project definition
- Communicating clear work scope
- Providing leadership for project collaboration
- Communicating clear business goals to design and construction team.

It's critical for clients to fully understand the importance of clear and timely communications. Contractors should not assume that clients bring to the table the communication skills necessary to avoid misunderstandings of project goals and priorities. Both contractor and client need to take an active role to remedy gaps in the communications process so that the project will be successful for everyone involved.

Box 8.2   Poor Performance or Poor Communications?

A general contractor was awarded a $20 million project to renovate a store for a major retail client. After 18 months, the client dismissed the contractor on the basis of missed milestones. The decision halted renovation for five months, including the holiday shopping season. The contractor sued the client, claiming that the managers undermined the project with numerous changes to its scope, slow responses to questions posed by the contractor, routine denials of change orders, and interference with subcontractors.

Although there was no evidence presented of any unethical behavior, it's easy to see how poor communication might be used to justify unethical actions by either the client or the contractor. Did the client deliberately undermine the project to justify terminating the contract? Did the contractor blame the client to cover up for poor performance? This is one contractor/client relationship that will most likely never have another chance.

Applicable Regulations and Standards

Earlier in this chapter, we discussed some of the ethical implications of gift giving and overbilling. These actions have even more serious consequences if they cross over the line from unethical to illegal, as in cases of bribery and false claims. There are a number of statutes that address these illegal actions.

Federal Bribery Statute

Bribery of public officials is addressed in Title 18, Section 201 of the U.S. Code. The law specifically prohibits anyone from giving or promising anything of value to any public official, either directly or indirectly, with the intent to influence that official to act in violation of his or her public duty. It also prohibits public officials from seeking something of value in return for influence.

Notice that the law uses the word, "intent." It's not necessary for a public official to accept the gift. Any gift given with the "intent" to influence that official may be considered a bribe. Because the law is so strict in its language, contractors working with any public official or representative, even if that person represents a public entity other than the federal government, must be aware of actions that may even appear to involve improper gift giving.

False Claims Act

False claims against the U.S. government are addressed in Title 31, Section 3729 of the U.S. Code. The law prohibits knowingly presenting false or fraudulent claims for payment or approval.

Actions that can be prosecuted under the False Claims Act include mischarging labor hours and using excessive unit prices. Other common aggressive billing practices include charging for hours not worked, invoicing for documents not produced, and misrepresenting the need for additional services. On the owner's side, clients can be prosecuted for certifying payments for contractors who have submitted inflated bills and approving work that is not completed.

Box 8.3 False Claims: Morse Diesel Int'l v. United
States

Between 1994 and 1995, AMEC Construction Management Inc.
(ACMI), which was known as Morse Diesel International at the time,
was awarded four federal construction contracts: two related to the
Eagleton Courthouse in St. Louis, MO; one for construction of the U.S.
courthouse and federal building in Sacramento, Calif.; and a fourth for
renovations to the U.S. customs house in San Francisco. Morse Diesel
submitted false bond reimbursement claims to the U.S. Government.
According to counterclaims filed in the U.S. Court of Federal Claims,
in addition to submitting the false bond reimbursement claims,
ACMI's parent company also received kickbacks from the company's
bond broker. After a lengthy litigation process, the United States
recovered more than $19 million from ACMI to resolve the allega-
tions of fraud, false claims, and kickbacks. The United States sought
damages and penalties under the False Claims Act, the Anti-Kickback
Act, and common law theories.

http://www.justice.gov/opa/pr/2009/February/09-civ-085.html

Anti-Kickback Act

The federal Anti-Kickback Act of 1986 is contained in Title 41, Section
51–58 of the U.S. Code. A kickback is compensation of any kind directly
or indirectly accepted by a public official from a vendor, contractor, or
subcontractor for the purpose of influencing the award of a contract.
Kickbacks can include money, fees, commissions, gifts, or gratuities. We
can differentiate between a bribe and a kickback by looking at examples.
When a contractor gives a gift to a public official with the intent to
influence the official's decision in the contractor's favor, that's a bribe.
When a public official makes a decision in a contractor's favor, and then
receives a portion, percentage, or cut of the monetary or other benefit,
that's a kickback.

Construction Participant Perspectives

One way that construction companies relay the importance of ethical client relationships is by including the client in their statements of core values. The global architecture and engineering firm AECOM accomplishes this clearly on the corporate webpage outlining the company's purpose and core values:

> AECOM sets the industry standard for client service. We are passionate about solving clients' problems and exploring new opportunities with them.

This sends the message to clients that their project needs are of prime importance to AECOM. It also lays the groundwork for establishing a corporate culture of client service.

Engineering firm CH2M Hill also puts clients into their company mission, vision and values. In addition, their "Little Yellow Book," written by the company's co-founder and available for free download from their website, lists simple maxims for ethical behavior in the workplace, including:

> The client is king ... or queen. Each one is important. An extra phone call or meeting him/her at the airport are the kinds of things which, in addition to outstanding work in their behalf, can keep clients in our camp.

For their part, clients can create an incentive for contractors by requiring bidders to have and abide by an ethics code to compete for projects. The Federal Acquisition Regulations already require that contractors for the federal government must have established a code of business ethics and conduct, a compliance training program, and procedures for reporting and taking corrective action.

Questions and Scenarios for Discussion

Evaluate the following scenarios.

1 You are out to dinner with your spouse at your favorite restaurant. One of your clients happens to be in the restaurant with a small party at another table. You ask the waiter to offer that table a bottle of wine that you will pay for. They accept your offer.

- Would the cost of the bottle of wine become an issue?
- Does it make a difference if the client is from a private company or a public agency?
- How would the situation be different if you offered to pick up the tab for the entire group?

2 You spot an omission in a bid. A portion of the work that you know must be done has been left out of the specifications. You put money in your bid to cover the cost of the work. Now the job has been awarded to your company.

- Should you notify the client about the mistake at no cost to them?
- Should you notify the client and then pursue a change order to cover the cost even though you have already included it in your bid?
- Would your decision be different if you discovered you had underbid a different item of work and you could use the client error to make up the difference?

3 Your company is a government contractor working on a federal construction project. Your project is going well and you are planning a holiday gala celebration with a no host bar. There is a large guest list that includes your employees, subs, vendors, spouses, and several government representatives. The event will include a drawing for prizes that have been donated by your company and your subcontractors. The winner is drawn randomly from the names of all the guests. Most of the prizes are valued at less than $400 and include power tools, electronics, sporting goods, apparel, and a liquor basket.

- Are there any ethics or compliance issues involved if the winner of the drawing is a federal employee?

- After the party, there are several unclaimed T-shirts and coffee mugs with your company's logo on them. Is there any problem giving these out to any of the remaining guests?
- What issues can you imagine if your company was hosting the bar?
- How might you organize this event to avoid any negative ethics and compliance issues?

References

CH2M Hill. Little Yellow Book. http://www.ch2m.com/corporate/about_us/business_ethics.asp

Ethisphere Institute: World's Most Ethical Companies. http://ethisphere.com/rankings-and-ratings/

Federal Acquisition Regulation. https://www.acquisition.gov/far/

FMI / CMAA (2005). 6th Annual Survey of Owners.

James, R. (2009). The Integrity Chain. 3rd ed. FMI Corporation, Raleigh, NC.

U.S. Code. Cornell University Law School Legal Information Institute. http://www.law.cornell.edu/uscode/text

Note

1   Contractor gratuities offered to Government personnel are subject to the restriction under the Standards of Ethical Conduct for Employees of the Executive Branch, 5 CFR part 2635.

9 Ethics and the Environment

Learning Objectives

After reading this chapter, you should be able to:

- Describe the concept of environmental justice.
- Explain what is meant by the "precautionary principle" and the "tragedy of the commons."
- Imagine and evaluate potential consequences to the environment resulting from common construction site management practices.
- Recognize the federal statutes that regulate environmental impacts from construction activities.

Introduction

Most of the focus of this text has been on ethical decision-making with respect to impacts on other people – clients, employees, subcontractors, vendors, etc. Environmental ethics asks us to evaluate impacts that may occur beyond the arena of our human relations. While it is natural to view our activities from a human-centered viewpoint, we must also keep in mind that all human activity is made possible by the natural resources of our planet. Negative environmental impacts can have both immediate and long-term consequences, affecting our personal, individual health; the quality of life that we enjoy; and our ultimate survivability as a species.

When we make decisions that impact a natural resource that is shared, such as clean air and water, there may be ethical consequences of those decisions, even if they are in strict compliance with laws and regulations. The concept of environmental justice has its roots in the belief that each individual has a fundamental right to a healthy environment. Historically, that right has been most frequently denied to poor people, and to minority communities. Although the federal government made strides to address racial discrimination in the 1960s with the Civil Rights Act, it wasn't until the mid 1980s that awareness was raised about another form of discrimination that was denying equal access to a pollution-free, healthy environment in which to live.

In 1982, in Warren County, North Carolina, a group of African American citizens and supporters came together to protest the dumping of contaminated soil into a hazardous waste landfill that the state had located and constructed in their community. This act of civil disobedience, which led to the arrests of more than 500 people, sparked a study by the U.S. General Accounting Office (GAO) of the racial and socioeconomic profile of communities adjacent to four landfills in the southeast U.S. The study found that three of the four communities were predominantly African American.

A larger study commissioned by the United Church of Christ (UCC) in 1986 found that "racial and ethnic Americans are far more likely to be unknowing victims of exposure" to hazardous wastes.[1] Whereas wealthier communities are better equipped to take a stand against the siting of a hazardous or industrial facility in their communities – the "Not In My

Backyard" or NIMBY position – poor communities have less access to information, and spend more time coping with the demands of daily life. They typically have neither the time, the transportation, nor the financial resources to advocate on their own behalf. In addition, industrial facility sitings often take advantage of the need in the local community for employment. The ability to earn a paycheck today can understandably take precedence over longer term health impacts. The UCC study found a consistent national pattern of the siting of commercial hazardous waste facilities in neighborhoods with high minority populations, with three of the five largest commercial hazardous waste landfills located in African American or Hispanic communities.

Beyond the U.S., the idea of environmental justice has been concerned with the exploitation of developing countries by multinational corporations who wish to take advantage of poorly enforced, lax, or non-existent environmental regulations for resource extraction, manufacturing, or waste disposal. According to the U.S. Environmental Protection Agency (EPA):

> Environmental Justice is the fair treatment and meaningful involvement of all people regardless of race, color, national origin, or income with respect to the development, implementation, and enforcement of environmental laws, regulations, and policies.[2]

The concept of environmental justice asks us to consider whether our decisions are being influenced by any preconceived ideas or beliefs we may have about the relative importance or value of the stakeholders.

Issues of environmental justice can arise during a construction project. If not managed properly, construction activities can be major sources of air, water, soil, noise, and even light pollution, negatively impacting the environment and the quality of life for workers on the site, and residents in neighboring communities. Construction project owners and managers must take care that equal attention is given to the environmental needs of the surrounding community regardless of racial or economic features or status. One method of evaluating potential impacts is to employ the *precautionary principle*.

In 1998, a group of scientific researchers gathered to discuss what they had become accustomed to referring to as "the precautionary principle." One outcome of the meeting was the 1998 *Wingspread Statement on the Precautionary Principle*, which summarizes the principle this way:

> When an activity raises threats of harm to the environment or human health, precautionary measures should be taken even if some cause and effect relationships are not fully established scientifically.[3]

The precautionary principle is a more scientifically defined expression of the old advice that it is "better to be safe than sorry." With respect to the environment, the ethical assumption behind the precautionary principle is that humans are responsible for the protection and preservation of the earth's natural resources, which are vital to the common good. It shifts the burden of proof from the potential victims of a harmful action to those who would implement the action. The Occupational Safety and Health Act is precautionary legislation that places the burden of responsibility on employers to provide a safe workplace. Precaution is often considered a moral virtue and one of the bases for sound judgment and decision-making.

When many people share a common resource, such as the earth's atmosphere, minerals, trees, or the fish in the ocean, there is a risk of severely depleting that resource if each individual believes he or she is entitled to an amount deemed necessary for his or her own short term personal gain or best interest. In the absence of regulations that place limits on how much each person can take for their own benefit, there is no incentive or motivation for individuals to preserve the resource for the common good. This phenomenon is referred to as the "tragedy of the commons." Many environmental regulations that are in place today attempt to prevent the tragedy of the commons and subsequent resource depletion.

Introductory Case Studies

Two construction company managers were convicted of mishandling asbestos cement pipe during an upgrade of a municipal water and sewer system and sentenced to six months in prison. Both men pled guilty to violating the asbestos work practice standards of the Clean Air Act. Federal investigators found that company workers, who were insufficiently trained and supervised, crushed the pipe during removal, releasing the asbestos and creating a health hazard. The material was then improperly disposed of on the construction site as well as at sixteen additional sites in the vicinity. The U.S. EPA ended up spending almost $4 million to clean up the disposal sites, and imposed penalties on the company to recover the cost.

Another firm that had five of its own untrained and unprotected employees remove asbestos from pipes and boilers inside a factory was fined $1.2 million for 27 willful and serious safety violations. The U.S. Occupational Safety and Health Administration said the fine was one of the largest against any employer. The state environmental protection agency learned of the removal when asbestos-covered pipe was sent to a recycling center. Among other things, the firm was cited for not monitoring airborne asbestos, failure to use particulate air vacuums for dust control, and improper asbestos waste disposal.

Considering these two case studies, and outside of the legal issues, whose rights to a healthy environment may have been harmed by the actions involved? Whose rights are being protected by the state and federal environmental protection agencies and the Occupational Safety and Health Administration? Should these rights take precedence over the right of a company to make a profit on a project? Would your answer be the same if either company qualified as a small, disadvantaged business?

Ethical Challenges

In the following sections, we will examine some of the common ethical challenges encountered by construction companies with respect to environmental protection.

Ethical Challenge: Impacts of Construction Site Activities

Construction site activities frequently require the removal of vegetation and topsoil and the subsequent exposure of large areas of unprotected soil. When rain from storm events falls onto the unprotected soil, some water will infiltrate into the ground, but much will run off, following the path of least resistance, to the nearest storm sewer or stream, carrying with it a burden of loose soil and sediment. The release of high levels of sediment into storm sewer outlets, or other receiving bodies of water has a harmful effect on water quality and on wildlife. The U.S. EPA considers storm water runoff to be one of the most significant sources of water pollution in the nation, and takes compliance very seriously, as is illustrated in Box 9.1.

The requirement for preventing soil erosion and the transport of sediment in storm water runoff from a construction site is a good illustration of the precautionary principle in action. It requires preventive steps to be taken to eliminate potentially polluted storm water runoff at the source, rather than waiting to see if any damage is caused to the waterway, and then reacting with expensive and most likely ineffective treatment.

Many contractors still view erosion and sedimentation control as a costly imposition. However, it's important for both contractors and clients to understand that costs associated with water quality protection must legally and ethically be considered as one of the basic costs of construction, just like materials, equipment, and labor. If construction projects do not take responsibility for protecting nearby water quality from the negative impacts of site activities, some other individual, agency, or municipality will end up paying for that protection or treatment. Costs that are created by an industry, but are paid by people or a community outside of that industry are referred to as externalities. Ethically, it is unjust to impose external costs on people or a community that do not share in the profits of the activity. The cost of environmental cleanup is one of the most common and overlooked categories of externalized costs.

Construction projects also have ethical responsibility to bear the cost of mitigation and control of other forms of pollution associated with site activities, including dust and noise. Dust can become airborne from temporary and haul roads, material handling, stockpiles, and site earthwork activities. Soil tracked out onto roadways by site vehicles can get washed into storm sewers and streams by rain, and is considered an element of construction site runoff. Noise can range from fairly unobtrusive power tools, to persistent reversing vehicle alarms, to extremely loud impacts from pile driving. Many impacts are an unavoidable part of the construction process,

Box 9.1   Erosion and Sedimentation Control: Wal-Mart
Clean Water Act Settlement

In May 2004, the Department of Justice and the U.S. Environmental
Protection Agency, along with the U.S. Attorney's office for the
District of Delaware and the States of Tennessee and Utah, reached
a Clean Water Act settlement for storm water violations at Wal-Mart
store construction sites across the country. Under this Clean Water
Act settlement, Wal-Mart agreed to pay a $3.1 million civil penalty
and reduce storm water runoff at its sites by instituting better control
measures.

In 2001, Wal-Mart settled claims that it had violated the storm water
requirements at about 17 sites across the country. That settlement
called for payment of a $1 million penalty and a compliance and
training program.

After the settlement, follow-up inspections at 24 Wal-Mart stores
revealed that violations continued. Specifically, EPA and state
inspectors found:

- failure to obtain permits for some sites
- discharges of excessive sediment to sensitive water ways
- failure to install and/or maintain adequate sediment and erosion
  control devices
- failure to develop and/or implement a storm water pollution
  prevention plan
- failure to inspect sediment control devices to ensure adequacy and
  condition and proper operation
- failure to develop an adequate plan for controlling sediment and
  minimizing erosion.

The settlement was the largest civil penalty ever paid for violations of
the storm water regulations.

http://www2.epa.gov/enforcement/wal-mart-ii-clean-water-act-settlement

but it is still important to consider potential mitigation efforts. In many cases, construction project managers simply need to consider what they might ask for if they happened to find themselves close neighbors to their own project sites, and develop strategies for communications and control. When impacts from site traffic, noise, odors, intrusive night lighting, or other factors can't be eliminated, communities should at least be informed of the estimated degree and duration so that they can prepare and plan around it.

Ethical Challenge: Construction Waste and Hazardous Materials Management

The construction process can generate substantial amounts of waste: excavated soil, scrap materials, pallets and packaging, used oil and lubes, miscellaneous rags and safety gear, and ordinary household trash from jobsite trailers and lunch rooms (Table 9.1). Many of these materials, such as fuels, paints, and adhesives, are considered hazardous waste, and must be disposed of in accordance with regulatory requirements. Demolition waste is also frequently contaminated with hazardous materials including lead, asbestos, or soil containing spilled fuel, heavy metals, or other contaminants. Even non-hazardous waste materials can have a major impact on local communities, requiring transportation from the jobsite and large landfill areas for disposal. In 2003, the U.S. EPA estimated that the quantity of construction and demolition waste generated annually was almost 170 million tons, with 39 percent coming from residential and 61 percent from non-residential sources.[4] To put this into perspective, and based on the number of people in the U.S. in 2003, this amount of construction waste is equivalent to a generation rate of 3.2 pounds per person per day, compared to 4.45 pounds of municipal solid waste (household trash) per person per day. That volume of waste from construction and demolition activities effectively almost doubles the amount of landfill area needed for disposal when added to the amount of municipal solid waste already being landfilled.

As was discussed earlier in this chapter, the siting of landfills, waste transfer stations, and other storage or disposal sites often raises issues

Table 9.1 Typical Components of Construction and Demolition Wastes

Waste Component	Examples
Wood	Forming and framing lumber, stumps/trees, engineered wood
Drywall	Sheetrock (wallboard)
Metals	Pipes, rebar, flashing, wiring, framing
Plastics	Vinyl siding, doors, windows, flooring, pipes, packaging
Roofing	Asphalt, wood, slate, and tile shingles, roofing felt
Masonry	Cinder blocks, brick, masonry cement
Glass	Windows, mirrors, lights
Miscellaneous	Carpeting, fixtures, insulation, ceramic tiles
Cardboard	Packaging
Concrete	Foundations, driveways, sidewalks, floors, road surfaces (all concrete containing portland cement)
Asphalt Pavement	Sidewalks and road structures made with asphalt binder

Adapted from U.S EPA, Estimating 2003 Building-Related Construction and Demolition Materials Amounts, Table 1. http://www.epa.gov/waste/conserve/ imr/cdm/pubs/cd-meas.pdf

related to environmental justice. The construction industry therefore has a responsibility to be aware of the types and amount of waste their projects are generating, and the communities in which those wastes end up. While construction waste may be "out of sight, out of mind" once it leaves the jobsite, the impact doesn't disappear with the hauler. It merely gets transferred to someone else's backyard.

There are a number of ways that contractors can minimize the amount of waste that ends up in a landfill. These include:

- requesting suppliers to reduce unnecessary packaging;
- planning material use to reduce the amount of scrap;
- donating unused but usable materials to non-profit salvage or resale outlets; and
- recycling materials when possible.

The preferred strategy for minimizing waste is source reduction – that is, avoid generating waste in the first place. The next preferred strategies are reuse and recycling of materials, both of which keep waste out of the landfill. Organizations such as the Construction & Demolition Recycling Association (CDRA) assist contractors in finding outlets for recyclable materials.[5] Many project owners, especially federal or state governments, now require contractors to have an approved construction waste management plan that estimates types and quantities of waste and identifies the disposition of those materials. A well-researched waste management plan can make financial sense. Landfill space is becoming increasingly expensive, with significant "tipping fees" imposed for every truckload, not to mention the fuel and equipment cost for the actual transportation of the waste. The money that is saved by avoiding landfill tipping (dumping) fees can offset additional costs that may be incurred by separating and transporting materials for reuse or recycling.

Ethical Challenge: Fuel Storage and Containment

Construction projects are dependent on fossil fuels to run vehicles and equipment. Depending on size and location, many projects must store substantial amounts of fuels and lubricants on the jobsite. Storage tanks and containers of petroleum products must be properly contained so that any spills do not come into contact with soil or water. Petroleum compounds range from volatile materials that readily evaporate at ambient temperatures, to heavier constituents that can remain in the soil for decades. Although regulations exist that govern the proper containment of fuel storage vessels, contractors often overlook individually minor but cumulatively significant spills that can occur in parking areas or where equipment is maintained.

As with erosion and sedimentation control, the cost of protecting the surrounding environment from contamination by spilled fuel products is a necessary cost of construction and should not be externalized to the community.

Applicable Regulations and Standards

There are a number of statutes that address the environmental impacts of construction. Table 9.2 provides an overview of some of the most common and relevant regulations.

Table 9.2   Environmental Regulations Related to Construction Activities

Activity or Impact	Regulations
Storm water runoff	**Clean Water Act** permit required for discharge of storm water runoff from the construction site. Storm water general permits are issued through the U.S. Environmental Protection Agency's (EPA) National Pollutant Discharge Elimination System (NPDES) program or the state NPDES permitting authority. Information on the storm water requirements: www.epa.gov/npdes/stormwater. Information about county and State storm water requirements through the Storm Water Resource Locator at www.envcap.org/swrl/
Dredged and fill materials Work affecting wetlands	**Clean Water Act Section 404** permit required for discharge of dredged or fill material to waters of the U.S. "Dredged" refers to material that is dredged or excavated from waters of the U.S., and "Fill" refers to material that replaces an aquatic area with dry land or changes the bottom elevation of a water body. Section 404 permits are issued by either the U.S. Army Corps of Engineers or, for certain waters, a state with an approved Section 404 permitting program. Information at: http://water.epa.gov/type/wetlands/
Construction and demolition wastes Hazardous wastes Wastes containing lead-based paint Fluorescent lamps containing mercury Petroleum product storage tanks	The **Resource Conservation and Recovery Act (RCRA)** lists hazardous wastes and their allowed concentrations, and also contains requirements for managing, treating, and disposing of hazardous wastes. Information at: http://www.epa.gov/epawaste/index.htm Lead information at http://www2.epa.gov/lead
Emissions from heavy-duty trucks Dust emissions	**Clean Air Act** requirements for mobile and stationary sources apply to some construction activities. Requirements are implemented primarily by states through their State Implementation Plans (SIPs). Information at www.epa.gov/oar/oaqps/ ▶

Activity or Impact	Regulations
Asbestos	The **Clean Air Act's National Emission Standard for Hazardous Air Pollutants (NESHAP)** for asbestos has to be met if there is a release of a Regulated Asbestos-Containing Material during demolition or renovation. Information at http://www2.epa.gov/asbestos

Adapted from U.S. EPA Federal Environmental Requirements for Construction.
http://www.epa.gov/compliance/resources/publications/assistance/sectors/
fedenvconstruction.pdf

Construction Participant Perspectives

In 2006, the Associated General Contractors of America (AGC) released its Environmental Agenda: a "commitment to encouraging and facilitating further improvement in the environmental performance of the construction industry."[6] The AGC developed the seven goals of its environmental agenda to help construction companies comply with environmental regulations, but also to encourage the industry to seek ways to reduce the environmental impacts of construction activities. In its environmental agenda, AGC challenges itself to:

- encourage environmental stewardship through education, awareness and outreach;
- recognize environmentally responsible construction;
- strengthen government support for positive incentives for environmental excellence;
- improve coordination and use of federal environmental rules, programs and efforts;
- provide contractors with tools to efficiently manage environmental exposures and risks of liability;
- identify opportunities to reduce the impact that construction means and methods have on the environment; and,
- identify ways to measure and report environmental trends and performance indicators of such trends.

AGC's Environmental Forum is a networking resource that allows members to stay current on environmental issues related to construction, participate in policymaking, and share best practices for limiting liability and reducing environmental impacts.

In the UK, the Considerate Constructors Scheme (CCS) is a nationwide initiative focused on improving the image of the construction industry. Among the focus areas, the CCS is particularly interested in reducing negative impacts of construction activities on the environment and the community. Construction projects that voluntarily register with the CCS agree to comply with five fundamental expectations of the Code of Considerate Practice.[7]

- Constructors should ensure sites appear professional and well managed.
- Constructors should give utmost consideration to their impact on neighbours and the public.
- Constructors should protect and enhance the environment.
- Constructors should attain the highest levels of safety performance.
- Constructors should provide a supportive and caring working environment.

The Code of Considerate Practice amounts to a code of ethical conduct for an individual construction project site. Registered projects are visited by monitors who review site practices and make recommendations if the project is out of compliance with the code. The CCS website provides a reporting avenue for members of the public who have witnessed CCS code violations, or feel they have been negatively impacted by a project. If an employee, neighbor, or other stakeholder submits a complaint about a registered project, the CCS acts as a mediator between the complainant and the relevant site or company manager until the issue is resolved. Rather than being a burden on participating construction firms, success stories posted on the CCS website describe how registering a project and complying with the code has helped companies take a fresh look at project plans and procedures. In addition, project managers describe how attention to the fundamental guidelines of the code assisted them in establishing a track record as a professionally run company, and resulted in new clients and more contracts. The independent review process helped companies to measure their practices against an industry benchmark allowing them to improve public perception and increase client satisfaction.

Questions and Scenarios for Discussion

Evaluate the following scenarios.

1   Your jobsite is going to be visited tomorrow by a group of managers from the home office, including the company's environmental compliance manager. You do a quick site inspection to make sure the fuel storage area is clean, but you notice that there are some oily stains on the ground in the equipment maintenance area. The last thing you need right now is some unnecessary attention on a couple of insignificant drips, so you have the laborers spread a few wheelbarrows full of fresh gravel over the stained area.

   • Are there any consequences associated with covering up a small fuel spill area?
   • What would your answer be if the environmental compliance manager noticed the fresh gravel and asked you about it?
   • What might happen if the environmental compliance manager asked the laborers about the fresh gravel?
   • What might be some other options for dealing with the spilled fuel?

2   You are managing a large-scale mixed use development project. You are currently working on the site preparation phase and there is a fairly large exposed area where the topsoil has been removed. That night, there is a heavy rain, and when you arrive in the morning, there is standing, muddy water in one of the low-lying areas. Fortunately, your other runoff protection measures have prevented water from leaving the site, but you need to keep working, so you arrange for a vacuum truck to remove the water from the site and empty it into a nearby storm drain.

- Are there any consequences associated with discharging the muddy water into the storm drain?
- Are there any circumstances when it might be permissible to discharge site runoff into a storm drain?
- What might be some other options for managing the accumulated water?

3 You have several containers of used oil from your project's vehicles and equipment. The local waste transfer station accepts used oil, so you ask one of your laborers to drop it off on her way home. When the laborer arrives at the transfer station, she finds out it's closed for the day and the gate is locked, so she leaves the containers right outside the gate and goes home.

- Are there any consequences associated with leaving the containers of used oil outside the gate to the transfer station?
- Do you see any issues with making the laborer responsible for the oil disposal?
- What might be some other options for disposing of the used oil?

4 You are working on a waterfront development project where you are required to drive piles with an impact hammer. The noise from the pile driving is extremely loud and disruptive. The project is located next to the city's busy financial district.

- What are your responsibilities to the nearby businesses?
- What would your responsibilities be if your project was located in an industrialized area next to a low-income housing project?

References

U.S. Environmental Protection Agency. Estimating 2003 Building-Related Construction and Demolition Materials Amounts. http://www.epa.gov/waste/conserve/imr/cdm/pubs/cd-meas.pdf

U.S. Environmental Protection Agency. National Pollutant Discharge Elimination Program. http://cfpub.epa.gov/npdes/

Notes

1 United Church of Christ, Commission for Racial Justice (1987). Toxic Waste in the United States. www.ucc.org/about-us/archives/pdfs/toxwrace87.pdf
2 http://www.epa.gov/compliance/ej/
3 Science and Environmental Health Network. *Precautionary Principle – FAQs.* http://www.sehn.org/ppfaqs.html
4 U.S EPA, Estimating 2003 Building-Related Construction and Demolition Materials Amounts. http://www.epa.gov/waste/conserve/imr/cdm/pubs/cd-meas.pdf
5 http://www.cdrecycling.org/
6 AGC's *Environmental Observer*, Issue 3-06, December 29, 2006.
7 Considerate Constructors Scheme, http://www.ccscheme.org.uk.

10 International Construction

Learning Objectives

After reading this chapter, you should be able to:

- Describe the ethical challenges that can occur when executing a construction project in a foreign country.
- Describe the basic requirements of the Foreign Corrupt Practices Act of 1977 and its impact on the conduct of international business.

Introduction

In this chapter, we will examine some of the ethical situations that one may encounter in the procurement and execution of construction projects in a foreign country. The standards of ethical behavior in the country may differ from those normally accepted in the United States, and U.S. construction companies need to understand how business is transacted in the country before deciding to pursue business in that country. In addition, some business practices that are acceptable in other countries may actually be unlawful for U.S. companies to adopt. International construction projects tend to present greater risk than those experienced with domestic projects. Construction firms considering entering the international construction market must understand the risks to be faced and adopt strategies to manage each risk. However, the opportunity for greater profits makes international construction attractive for those construction firms that are able to manage their risk exposure.

Project owners for international construction projects can be grouped into five general categories:

- United States government agencies
- multinational corporations
- international government agencies
- local government agencies
- local companies.

Each of these project owners presents different levels of risk. Contracts with U.S. government agencies will comply with the Federal Acquisition Regulation. Contracts with multinational corporations and international government agencies tend to be well-known international documents, whereas contract provisions used by local government agencies and local corporations may not be well known to U.S. construction firms. Irrespective of the project owner, labor, materials, and subcontractors usually are procured locally, and construction contractors must comply with the foreign nation's regulations and procedures. Because non-U.S. construction firms may not be constrained by U.S. laws, it may be tempting to adopt local ethical standards. Such behavior may lead to the foreign country terminating the construction company's ability to conduct business in the country and/or may subject the company to prosecution in the United States.

Introductory Case Study

Acme Construction was interested in obtaining a contract for the construction of a major hotel in the capital city of a foreign country and selected a project manager to investigate the situation and develop a tender for the project. The project manager requested the solicitation documents for the project and carefully reviewed them. He noticed that the contract documents to be used on the project were in a foreign language as were the solicitation documents. Since he was unable to read the documents, he hired an interpreter to translate the documents. The project manager carefully reviewed the translated documents and determined that the design of the hotel did not seem complete, but that the project owner required a lump sum tender or bid for its construction. The project location was described, but limited soils information was provided.

To obtain a better perspective of the local business environment and inspect the proposed construction site, the project manager traveled to the capital city. While there, he met with the project owner's representative to discuss the project and visit the project site. He learned from the project owner's representative that some of the construction materials needed for the project were not available in the country and must be imported. He learned that the procedures for obtaining customs clearance for imported materials could take significant time. He also determined that a construction firm must determine that sufficient local labor is not available before being allowed to import skilled labor and that any local labor to be hired must be obtained from local contractors. The availability of experienced subcontractors was quite limited, and the project manager would need to hire local supervisors to manage local construction workers, because few of the workers speak other than the local language. The project owner's representative informed the project manager that he could ensure that Acme Construction would win the construction contract for the hotel if the company would compensate him for his efforts.

After investigating the situation in the foreign country, the project manager returned to the United States to prepare Acme's bid or tender for the project. Because the hotel design was not complete and some aspects of the site conditions were unknown, he decided to add a 30 percent contingency to the bid. Was the inclusion of the contingency ethical or legal? The project manager also decided to include a fee for the project owner's representative to ensure award of the contract and a fee for the local customs agent to ensure timely process of imported materials. Were these fees ethical or legal?

Ethical Challenges

The following sections explore some of the unique ethical challenges that can arise for international construction projects.

Ethical Challenge: Gift Giving

In some foreign countries, it is common business practice to give gifts to project owners or their agents to obtain business and to government officials to obtain expedited processing of government transactions. Such practices are considered unethical, and perhaps unlawful, in the United States. It is easy to rationalize conducting business in the same manner to be competitive with other international construction firms, but it is not a good business strategy. Many construction firms, however, refuse to do business in countries where such practices are common and choose to select countries with ethical business practices.

Ethical Challenge: Construction Standards

Many foreign countries have not adopted the types of rigorous building codes that ensure that completed projects provide sufficient life safety measures for occupants. Local building officials may not always require or perform detailed inspections of completed projects. In such situations, it is unethical, and perhaps unlawful, for construction companies to use substandard materials or take short cuts in constructing the project. Ethical construction firms will fully comply with contract requirements, even if the work is not inspected by the owner's representatives or local building officials. The reputation of the construction firm is greatly impacted by the quality of completed projects.

Ethical Challenge: Environmental Standards

Many foreign countries may not have or enforce sufficient or rigorous environmental regulations and may provide no guidance on the proper disposal of hazardous wastes. As discussed in Chapter 9, most construction projects generate some wastes that are considered hazardous, such as paints or solvents, and they must be disposed of properly. Even if the country in which a project is being constructed does not have environmental regulations or standards, it is incumbent on the construction firm to determine safe means for control and proper disposal for all hazardous waste. Failure to do so represents unethical behavior.

Ethical Challenge: Project Safety

Many foreign countries do not have comprehensive workplace safety and health requirements. In some locations, project safety is ignored by workers and management in the performance of construction operations, either deliberately, to expedite the work, or out of ignorance or lack of training. It is unethical for a U.S. construction firm not to require the same level of jobsite safety as required for a U.S. project. Proper safety equipment should be provided, and workers required and properly trained to use it. International construction firms should make jobsite safety a priority to demonstrate concern for the labor force as well as eliminate the costs of accidents. Failure to enforce good safety practices is unethical.

Ethical Challenge: Work Hours

Construction workers on many international construction projects may have been recruited outside of the country in which the project is being constructed. These workers tend to be housed in temporary project support facilities for the duration of their employment on the project. Since they are separated from their families, many of them want to work seven days per week for long hours in order to maximize their income. In some foreign countries, such practices are not allowed by local employment regulations. In other countries, there may be no prohibition to adopting such working conditions. Workers without adequate rest are susceptible to accidents and may not pay close attention to detail in their work. International construction companies should carefully monitor the number of hours worked to ensure that workers receive adequate rest. In most cases, construction workers should not be allowed to work more than six days per week. Failure to enforce procedures that focus on worker welfare is considered unethical.

Applicable Regulations and Standards

In response to investigations conducted in the 1970s, the U.S. Congress passed the Foreign Corrupt Practices Act (FCPA) in 1977 making it unlawful to bribe foreign government officials to obtain or retain business. The abuses identified in the 1970s ranged from bribery of high foreign officials in order to secure some type of favorable action by a foreign government to facilitating payments that allegedly were made to ensure that government employees discharged certain responsibilities. The FCPA makes it unlawful for a company (as well as for any employee or agent) to offer, pay, or promise to pay anything of value to a foreign official for the purpose of inducing the recipient to misuse his or her official position to wrongfully direct business to the payer.

The FCPA differentiates between corrupt payments, or bribes, and payments that are necessary to facilitate the implementation of the project within the context of the local culture. This can seem like a tricky distinction since facilitating payments are not acceptable in the U.S., but may be standard in other countries to ensure "routine government action" by a foreign official. The following actions are listed as those that may require facilitating payments:

- obtaining permits, licenses, or other official documents
- processing governmental papers, such as visas and work orders
- providing police protection and mail service
- providing telephone service, utility services, and loading and unloading services
- scheduling of inspections associated with contract performance or transit of goods.

Companies that enter the international construction market must ensure that its employees and agents are fully aware of the provisions of FCPA.

Applicable ethical standards are to practice good faith and fair dealing with project owners and in the selection of material suppliers, subcontractors, and life support contractors. Any construction company planning to enter the international construction market needs to obtain legal counsel who are familiar with the contracting procedures and legal requirements of all countries in which the company plans to pursue business. It is crucial to understand the risks associated with procurement and contracting actions and the legal framework for resolving disputes.

Construction Participant Perspectives

To be successful in international construction, construction firms must understand the contracting and business practices of project owners as well as the country in which the project is located. They must fully understand the requirements of the FCPA to ensure that none of their employees or agents engages in an unlawful behavior. Company leaders must ensure that individuals selected to lead the project teams fully understand company policy for ethical behavior. In most cases, when a construction firm decides to enter the construction market in a foreign country, they wish to receive future projects when their initial project is completed. If this is the company's business strategy, the company must develop a reputation for ethical behavior and ensure that all employees understand their role in developing such a reputation.

Questions and Scenarios for Discussion

Evaluate the following scenarios.

1 International Constructors, a U.S. firm, is interested in obtaining the contract for the construction of a major industrial facility to be located in a foreign country. The contractor's project manager requested the solicitation documents and investigated the procedure that was to be used for the selection of the construction contractor for the project. He determined that the mayor of the city in which the project is to be

constructed was on the Board of Directors of the company planning to build the facility. Wanting to ensure receipt of the construction contract, International Constructors' project manager invited the mayor to lunch and promised to make a large contribution to the mayor's re-election campaign if International Constructors received the construction contract.

- Were the actions of International Constructors' project manager ethical? Were they lawful?
- How would you have handled the situation?

2　Continental Construction has a contract for the construction of a water treatment plant in a foreign country. The design documents were about 90 percent complete when the construction contract was awarded to Continental. During the construction of the project, Continental's superintendent noticed that the electrical design did not require installation of safety features typically required in the United States. The superintendent decided not to ask the project designer about the issue or require the installation of the safety features.

- Were the actions of the superintendent ethical?
- How would you have handled the situation?

3　National Construction Company has a contract for the construction of an oil refinery in a foreign country. The project is nearing completion, and the project superintendent discovered significant quantities of leftover hazardous materials. He asked the project manager what procedures should be followed in disposing of the hazardous materials. The project manager submitted requests for guidance to both the project owner and a local government agency. Both responded that there were no regulations related to the disposal of hazardous waste. The project manager then told the superintendent to dispose of the hazardous materials in the nearby local landfill.

- Were the actions of the project manager ethical?

- How would you have handled the situation?

4 Allied Construction has a contract for the construction of a power plant in a foreign country. The workforce employed on the project contains workers from multiple countries. The project superintendent noticed that several of the steel workers working on the roof of the plant were not wearing hard hats or gloves. Since the country in which the plant is being constructed has no workplace safety regulations, the superintendent chose to ignore the lack of safety equipment.

- Were the actions of the superintendent ethical?
- How would you have handled the situation?

5 Cascade Constructors has a contract for the construction of a major hotel in the Middle East. Many of the workers employed on the project were imported from Asia. The workers requested that they be allowed to work ten hours per day seven days per week in order to make significant overtime wages. Cascade's project manager, seeing the benefit of early completion of the project, readily agreed.

- Were the actions of the project manager ethical?
- How would you have handled the situation?

Reference

For more information on the Foreign Corrupt Practices Act, please visit the Department of Justice website at http://www.justice.gov/criminal/fraud/fcpa/.

11 Emerging Topics

Learning Objectives

After reading this chapter, you should be able to:

- Explain the similarities and differences between ethics and compliance, and corporate social responsibility.
- Identify corporate social responsibility initiatives that might benefit a construction company.
- Give examples of ethical challenges associated with the use of collaborative digital information.
- Describe ways in which information from social media could help or hurt a company.

Introduction

As we have seen in the chapters leading up to this point, ethical issues faced by the construction industry, as for business in general, deal largely with matters pertaining to honesty and fairness in the performance of work, the rights of employees in the workplace, and personal and corporate conflicts of interest. As we move ahead into the twenty-first century, however, companies and individuals are facing new issues related to the use and prevalence of technology, and the responsibilities of corporate citizenship. In this chapter, we will explore some of the ways in which companies are beginning to deal with these emerging topics.

Introductory Case Study

A regional construction company has just started a blog to highlight its current projects. The blog, which is linked to the company website, features photos of different projects with brief descriptions. The company's webmaster and the marketing manager both have administrative rights to the blog site. The marketing manager thought it would help boost the company's reputation if the blog posts received some positive comments. She created a half dozen or so email accounts with different usernames and used them to post comments to the blog praising the company for their high quality work and dedication to the community. Shortly afterwards, the company was contacted by two new clients inviting the company to bid on upcoming work. Is this an ethical marketing strategy?

Ethical Challenges

With advances in technology, and the prevalence and popularity of social networks, construction companies are challenged by an unprecedented need to manage their information and their reputations. In the following sections, we will look at the connection between company codes of conduct and corporate social responsibility, as well as ethical challenges related to the use of digital information and social media.

Ethics, Sustainability, and Corporate Social Responsibility

The concept of social responsibility grew out of the civil rights movements of the 1960s and 1970s. As ethics and compliance programs have become more common among most of the larger firms in the construction industry, a related focus on corporate social responsibility has also taken hold. Whereas ethics and compliance programs are aimed at internal operations and avoidance of misconduct by employees, corporate social responsibility, or CSR, looks outward on the impact of the company on the communities in which the company's offices and projects are located.

Ethics and compliance and CSR programs both seek to achieve fair and ethical treatment of a company's stakeholders, but take different approaches. Company codes of conduct communicate the firm's shared values and offer guidance for decision-making in areas of primary risk, such as conflicts of interest, use of confidential information, and contract performance. CSR programs seek to define the firm as a "good neighbor" through efforts that include charitable giving, volunteer programs, public information events, and initiatives to select socially and environmentally responsible supply chains for materials and services.

CSR also encompasses sustainability reporting. There is an excellent definition of sustainability as an ethical imperative, drawn from the writings of St. Vincent DePaul, and published by the Institute for Business and Professional Ethics at DePaul University:

> Sustainability refers to personal and organizational commitment to provide the best outcomes for the human and natural environments both for immediate and future needs.[1]

Companies commonly incorporate sustainability initiatives into their corporate culture by measuring performance against three concerns: economic growth, environmental protection, and social equity. These three concerns comprise what is referred to as the triple bottom line. Ethical issues in business can arise in all three areas. There are various frameworks for sustainability reporting. The Global Reporting Initiative (GRI) has developed one such framework.

In GRI sustainability reporting, firms disclose various aspects of their company profile, government, and operations as they relate to the triple bottom line. A number of the standard disclosures are directly related to the area of ethics and compliance:

- describe the organization's values, principles, standards and norms of behavior such as codes of conduct and codes of ethics;
- report the internal and external mechanisms for seeking advice on ethical and lawful behavior, and matters related to organizational integrity, such as helplines or advice lines;
- report the internal and external mechanisms for reporting concerns about unethical or unlawful behavior, and matters related to organizational integrity, such as escalation through line management, whistleblowing mechanisms or hotlines.[2]

For publicly held companies, the dedication of financial and human resources to CSR and sustainability reporting can create an uncomfortable tension when confronted with the duty to maximize profits for shareholders. As a result, there has been a substantial amount of research directed towards identifying tangible and intangible benefits of CSR programs.

In a 2013 report released by the Boston College Center for Corporate Citizenship and Ernst & Young LLP, a table presented survey results showing ways that sustainability reporting provides value. The top four motivations given by the 579 respondents (including approximately 20 who worked for construction companies) were: transparency, competitive advantage, risk management, and stakeholder pressure. What's more, 50 percent of the respondents felt that sustainability reporting gave their company a competitive advantage.[3]

CSR both reflects and influences a company's culture of ethics. The Ethics Resource Center has reported that corporations that invest in CSR benefit from improved image and reputation, which in turn allows them to attract more customers and investors, and compete more effectively.[4] It's not hard to imagine how a positive image and relationship with the community could benefit a construction company. For example, a construction firm might face strenuous objections and possible delays for a project that involves traffic or noise impacts to a local community. The company is more likely to gain support and tolerance of the project if the firm has already demonstrated itself to be a good neighbor, cultured a positive relationship with the community and its residents through charitable and volunteer events, and been responsive to community concerns.

Because of the commonalities between ethics and CSR, a strong CSR program is viewed as an expression of a strong culture of ethics within a company. That is, a company's CSR initiatives gain validity when the firm's employees are known to behave ethically. In a similar manner, employees tend to behave more ethically when they work for a company with a strong commitment to CSR. Many aspects of CSR have their foundation in the various ethical theories including fairness, acting for the common good, and striving to accomplish the greatest good for the most people.

In May 2013, Engineering News Record reported on the status and impact of corporate giving by construction firms. Referring to a 2011 study by Forbes Insights and Hewlett-Packard, ENR noted that:

> Allowing employees to follow their passions and support causes they care about helps increase employee loyalty and keeps people motivated during times when salary bumps are scarce. The study says volunteering enriches creativity in product and market development, and skill-based philanthropic efforts showcase the company's talents and help attract new business. In construction, such projects might include building low-income housing, rehabbing homeless shelters or constructing playgrounds for area schools.[5]

For a deeper look into the evolution of corporate social responsibility, and related ethical theories, the chapter on *Social Responsibility and Business Ethics*, by Buchholz and Rosenthal is highly recommended. The complete reference is provided at the end of this chapter.

Ethics and Digital Information

With most company records now stored electronically, firms can be at great risk if sensitive or confidential information is accessed by

---

Box 11.1   Construction Companies Give Back to the
Community

Clark Construction Group, Bethesda, Maryland

The company's charitable foundation specializes in helping wounded,
active-duty service members and veterans from nearby military
hospitals.

Turner Construction, New York

Operates mentoring and training programs in communities where
employees live and work.

Saunders Construction, Centennial, Colorado

Launched a program called Building Confidence in Kids, which has
rebuilt a playground, refurbished rooms and built an on-site day care
center at an affordable housing community.

Sundt Construction, Tempe, Arizona

Employee and matching company donations support community
organizations in Arizona, California and Texas, where Sundt has
offices.

McCarthy Building Companies, St. Louis, Missouri

The company's Heart Hats initiative supports local projects that
benefit low-income families and children.

Source: *Corporate Giving Helps Communities and Improves Construction
Industry. ENR, May 2, 2013.*

---

unauthorized personnel, or by authorized personnel in violation of the
company's code of conduct. Technical design information, construction
methods and sequences, client lists, marketing strategies, and historical

job cost data are examples of confidential and proprietary information that give a company its competitive advantage. Employee records and human resource data that contain Social Security numbers, bank account numbers for direct deposit of paychecks, and other private data must also be safeguarded against exposure or theft. All employees should receive training to help them recognize sensitive information and know how to use it appropriately.

A first level of data protection is often the company's private intranet, accessible only by employee login. There may be different permission levels so that employees only have access to the data that they require to perform their jobs. Electronic data must be backed up securely and frequently, not only to protect against loss, but also to allow access to previous versions of documents that may have been inadvertently or inappropriately changed or edited. Employee training should include instruction on creating secure passwords and emphasize the importance of not sharing or revealing login or password information.

Digital information is both easily accessed and highly portable. Companies often rely on confidentiality agreements to set boundaries on what can and cannot be shared outside of the firm. Confidentiality or non-disclosure agreements are legal contracts that restrict people from sharing sensitive information to which they may have been granted access for business purposes. By signing such a document, an employee agrees to protect and not disclose the information described in the agreement. Employees have an ethical and legal obligation to comply with the requirements of any confidentiality or non-disclosure agreements by not intentionally revealing protected information. Non-disclosure requirements may extend to subcontractors and vendors who have access to proprietary project information. Sharing confidential information that is protected by a non-disclosure agreement is not only a violation of a legal contract; it is also a breach of accepted business ethics and a betrayal of trust.

In the construction industry, any competitive advantage can mean the difference between winning or losing a bid or proposal. The more you know about the way your competitors win projects, the better you can design your own winning strategy. The most obvious legitimate source of competitive information is that which is publicly available, including news articles and press releases, company newsletters and brochures, public filings, and information available on a company's public website. There is a simple rule of thumb to determine whether competitive information has been ethically obtained: if the means used to obtain the information involve lying, misrepresentation, or deception, the tactic is probably unethical.

Although a company may enjoy a short-term benefit by illegally or unethically obtaining information about a competitor, such practices never pay off in the long run. Stealing confidential information is not a successful long-term business strategy, and is no substitute for developing sound marketing and business development practices based on satisfied clients and a solid reputation for high quality performance.

Confidentiality agreements or contractual clauses are particularly important on projects that use a collaborative approach. A building information model (BIM) is an example of a collaborative effort that relies on massive data sharing. If a project will require the use of a model that is jointly created and shared, it is crucial for all parties involved to understand and agree upon who owns the BIM, who are the authorized users, and when each user's access to the BIM begins and ends. The American Institute of Architects offers a free BIM protocol document that addresses these issues (http://www.aia.org/).

When multiple parties participate in the planning and creation of the BIM, who is ultimately responsible for errors in design or code compliance? During the planning stage, if a contractor finds a design error that can potentially lead to a substantial change order in his or her favor, that contractor has an ethical obligation to communicate the error to the other parties. An ethical check would most certainly reveal that the contractor would not wish to be held responsible if he or she were the one who ended up paying for the mistake.

Ethical issues can also arise when multiple parties have access to and can make changes to a BIM if one party decides to make changes to their own advantage without communicating those changes to the other parties. For example, a mechanical contractor who decides to alter the configuration of piping or ductwork could negatively impact the performance of the electrical contractor if those alterations are not communicated and agreed upon ahead of time. The project's BIM protocol should anticipate such challenges and make sure all parties are aware of and committed to following appropriate procedures.

Ethics and Social Media

According to a 2012 survey by McKinsey, more than 1.5 billion people around the globe now have an account on a social networking website, and almost one in five online hours is spent on social networks – increasingly via mobile devices.[6] Even if social networking sites are blocked on the company network, employees can still gain access via smartphones or other personal devices.

In terms of percentage of the workforce, a 2012 study by the Ethics Resource Center revealed that 75 percent of U.S. employees belong to one or more social networking site.[7] The survey further showed that although some employees only connect to their social network during breaks, many others check in at least once an hour, or are constantly connected.

Unless an employee has the specific job function to use and monitor social media for the company, it's probably safe to assume that very little (if any) of the time spent by most workers on social networks during work hours is required to accomplish their jobs. Personal time spent on social networks might have nothing to do with the employer, or could include commentary about the employer, work projects, colleagues, or clients. This commentary could not only compromise company confidential information, but social networks also present a potential outlet for retaliation, and could inhibit workers' freedom to report workplace misconduct.

Companies themselves are increasingly using social media for marketing and for recruitment. Jobvite's 2013 Social Recruiting Survey of 1,600 recruiting and human resources professionals revealed that 94 percent of those surveyed either used, or planned to use, social media for recruiting.[8] The growing popularity of social media-based information comes with ample opportunity for unethical behavior including misrepresenting one's self or others or posting abusive, embarrassing or harassing comments, links or photos. Firms that rely on social media as a marketing tool need

---

Box 11.2   Engineering and Construction Firms Develop
Social Media Policies

Burns & McDonnell, Kansas City, Missouri

Burns & McDonnell has a formal social media policy and training
program for employees. The company's policy allows employees to
speak about the company, but not on its behalf; prohibits use of the
company's name or logo without first checking with the corporate
marketing department; and discourages personal use of social media
during work hours.

Barton Malow Company, Southfield, Michigan

Barton Malow's social media policy covers the legal liabilities of
sharing information online, disclosure requirements, practices for
company-appointed bloggers and guidance for all employees who may
decide to use personal social media outlets to discuss the company.

Source: *Writing the Social Media Policy Handbook at Burns & McDonnell, HOK*
ENR, October 18, 2011.

---

to have policies and procedures in place to help employees know what
is and is not appropriate to post and link to. Company social media sites
need to be carefully monitored and administered to prevent inappropriate
content. Firms that use social media in their recruitment efforts need to
verify posted credentials, and have a healthy suspicion of information that
appears either overly complimentary or unjustifiably negative.

So far, most companies do not perceive use of social networks by
employees to be a significant risk, although it has close ties to data privacy
and confidentiality, which companies almost universally identify as a
significant risk. Firms are just beginning to develop policies and training
about the use of social networks in the workplace, as well as addressing
social media use in the company code of conduct. Some companies are

taking the initiative to incorporate social media as a component of their ethics and compliance training as another way to engage employees.

With the rapidly changing landscape of social networks, company policies should focus more on general behavior guidelines and ethical use of social media and less on the features of specific networks.

In addressing social networking in a company code of conduct, employers might consider adding another question to their ethics checklist: "Would you feel comfortable if you read about this action in a post on a social network?" The ERC study showed that when companies had a policy on social networking, and made sure to communicate that policy to their employees, those employees were less likely to engage in online behavior that put their company at risk. Ultimately, training employees about appropriate use of social media is just another aspect of professional communication in general. From an ethical standpoint, social networking policies often refer to the golden rule – that is – employees should consider the potential consequences and impacts that a post or comment on a social network might have on the company or a coworker, and imagine being on the receiving end of that action.

Questions and Scenarios for Discussion

Evaluate the following scenarios:

1 A project engineer was accidentally copied on an email from a friend who works for one of the company's current clients. The email included cost detail plus owner analysis and comments on four contractors bidding on an unrelated project. The four contractors are all competitors of the project engineer's company.

- What action should the project engineer take?

2 Bob is working on a proposal for a new project. The proposal manager has given everyone on the team access to a private folder on the company server that contains confidential information to be used in the proposal. While looking through the contents of the folder, Bob finds a document that lists ten employees who have been included in

the proposal and their current hourly rates. Bob notices that one of the employees on the list is a coworker who has the same qualifications and less seniority than Bob, but makes almost 20 percent more money. Bob immediately feels angry. He prints out a copy of the file with all the rates on it, and stashes it in his desk. He thinks it will be useful for negotiation next time he is up for a salary increase. He also decides to take a little longer for lunch from now on. He figures if he's not getting paid as much as his coworker, he deserves to take a little more personal time.

- Is Bob's behavior justified?

3 A new administrative assistant in the HR department of a medium-sized construction firm was going through some old files and found some employment applications that were several years old. Since they were so old, he decided to put them into the recycle bin. There was so much other paper in there anyway, the old applications would just get buried. While the recycle bin was waiting to be picked up, a "dumpster diver" discovered the applications and used the information to steal the identity of two of the applicants.

- What action should the company take?

References

Buchholz, Rogene A. and Sandra B. Rosenthal (2002). Social responsibility and business ethics. *A Companion to Business Ethics*. Robert E. Blackwell Publishing, 2002. *Blackwell Reference Online*. (01 September 2013).

DePaul University Institute for Business & Professional Ethics (2007). Ethics 101: A Common Ethics Language for Dialog. Compiled by the Ethics Across the Curricula Committee. http://commerce.depaul.edu/ethics

Ernst & Young, (2013). Value of Sustainability Reporting. A study by the Center for Corporate Citizenship and Ernst & Young LLP. Complete report available at www.bcccc.net/pdf/valueofsustainability.pdf

Ethics Resource Center (2013). The National Business Ethics Survey of Social Networkers: New Risks and Opportunities at Work. www.ethic.org/nbes

Ethics Resource Center (2011). The Interplay between Ethics and Corporate Responsibility Opportunities & Challenges for Ethics Professionals – a white paper by the Ethics Resource Center Fellows Program, July 2011. www.ethics.org

McKinsey Global Institute (2012). The social economy: unlocking value and productivity through social technologies. www.mckinsey.com/insights/high_tech_telecoms_internet/the_social_economy

Notes

1 Ethics 101: A common ethics language for dialogue. Compiled by the Ethics across the Curricula Committee at De Paul University (2007), p. 9.

2 GRI – G4 Sustainability Reporting Guidelines: Reporting Principles and Standard Disclosures, General Standard Disclosure numbers G4-56, 57, and 58.

3 Ernst & Young, LLP – Value of Sustainability Reporting (2013).

4 ERC – The Interplay Between Ethics and Corporate Responsibility: Opportunities and Challenges for Ethics Professionals (2011).

5 Engineering News Record – Corporate Giving Helps Communities and Improves Construction Industry Image (05/02/2013).

6 McKinsey Global Institute – The social economy: unlocking value and productivity through social technologies (2012).

7 ERC – The National Business Ethics Survey of Social Networkers (2013).

8 Jobvite Social Recruiting Survey Results 2013. http://recruiting.jobvite.com/

Index